Physics and Mechanics of New Materials and their Application

Physics and Mechanics of New Materials and their Application

Yun-Hae Kim

Sanjay Kumar

National Korea Maritime and Ocean University, Korea

World Scientific

NEW JERSEY · LONDON · SINGAPORE · BEIJING · SHANGHAI · HONG KONG · TAIPEI · CHENNAI

Published by

World Scientific Publishing Co. Pte. Ltd.

5 Toh Tuck Link, Singapore 596224

USA office: 27 Warren Street, Suite 401-402, Hackensack, NJ 07601

UK office: 57 Shelton Street, Covent Garden, London WC2H 9HE

Library of Congress Control Number: 2024950796

British Library Cataloguing-in-Publication Data
A catalogue record for this book is available from the British Library.

PHYSICS AND MECHANICS OF NEW MATERIALS AND THEIR APPLICATION

ISBN 978-981-98-0771-0 (hardcover)
ISBN 978-981-98-0772-7 (ebook for institutions)
ISBN 978-981-98-0773-4 (ebook for individuals)

For any available supplementary material, please visit
https://www.worldscientific.com/worldscibooks/10.1142/14170#t=suppl

Desk Editor: Muhammad Ihsan Putra

Typeset by Stallion Press
Email: enquiries@stallionpress.com

https://doi.org/10.1142/9789819807727_fmatter

Foreword

The rapid advancements in physics, material science, and engineering over recent decades have revolutionized industries and technologies across the globe. As new materials emerge, their unique physical and mechanical properties present exciting possibilities for innovation in fields such as aerospace, electronics, energy, and biomedical engineering. This book, "Physics and Mechanics of New Materials and Their Application," seeks to provide a comprehensive overview of these emerging materials, with a focus on their underlying physics, mechanical behavior, and potential applications.

This volume is a collaborative effort by leading researchers and experts in material science, engineering, and applied physics. It is intended to serve as a resource for both researchers and professionals who are interested in understanding the latest developments in the field, as well as students seeking to deepen their knowledge of the science behind these cutting-edge materials.

The chapters of this book cover a range of topics, from the synthesis and characterization of advanced materials to their practical application in modern industries. Special attention is given to the mechanics of these materials, exploring how their structure, composition, and external conditions influence their performance in real-world applications. Case studies and examples are provided to illustrate how theoretical principles translate into practical solutions.

As editors and co-editors of this volume, we are deeply grateful to all the contributors who have shared their insights and expertise, and we hope this book will inspire further research and innovation in the exciting field of physics, material science, and engineering. It is our belief that the exploration of new materials will continue to unlock possibilities for technological advancements that will shape the future of various industries.

We hope this book proves to be a valuable resource and inspiration for your work.

Editors
Prof. Yun-Hae Kim and Dr. Sanjay Kumar
National Korea Maritime & Ocean University, Korea

Contents

Influence of rare earth on the microstructure and mechanical properties of Al–Zn alloy*

Bui Thi Ngoc Mai

Faculty of Ship Building, Vietnam Maritime University,
484 Lach Tray, Le Chan, Hai Phong, Vietnam
maibtn@vimaru.edu.vn

Nguyen Xuan Dong[†], Pham Mai Khanh[‡,¶] and Tran Duc Huy[§,¶]

School of Materials Science and Engineering,
Hanoi University of Science and Technology,
No. 1 Dai Co Viet, Hai Ba Trung, Hanoi, Vietnam
[†] *dongnx1686@gmail.com*
[‡] *khanh.phammai@hust.edu.vn*
[§] *huy.tranduc@hust.edu.vn*

This paper discusses the influence of Vietnam rare earth (RE) with the composition of 69.4% Ce and 30.5% La (modification content of 4% and 6%) on the microstructure and mechanical properties of Al–5Zn–3.5Mg–1.2Cu alloy when the temperature modification is changed at 750°C in 150, 200, 250 s. By optical microscopy, the results showed that the microstructure of the modified sample is finer than the nonmodified sample. By the SEM and EDS analysis, at the grain boundaries of this alloy, $MgZn_2$ and/or $Mg_3Zn_3Al_2$ intermetallic phases appeared. The modified elements (La, Ce) in RE and aluminum were formed in the intermetallic $Al_{11}RE_3$ and Al_2RE phases, which prevents the development of the dendritic α phase. The result showed Al–5Zn–3.5Mg–1.2Cu alloy modified with 4% RE in 150 s which is the most optimal result, the grain size is 6.67 μm and the hardness is 107 HB.

Keywords: Rare earth; intermetallic phase; grain size; modification.

[¶]Corresponding authors.

*To cite this article, please refer to its earlier version published in the *International Journal of Modern Physics B*, Volume 34, 2040125 (2020), DOI: 10.1142/S0217979220401256.

1. Introduction

Al–Zn–Mg–Cu alloys which have the highest strength in aluminum alloys can be deformed and heated. After aging, this alloy has a high yield limit which is smaller than the strength limits by about 20–30 MPa and higher than the Dura D16 by about 40–50%. Therefore, this alloy is used for products that require high deformation and still retain certain strength.[1,2] There are a lot of researches on the modification of aluminum and zinc alloys in the literature. For example, Tian and his colleagues studied aluminum modification by adding a small amount of Sc and Zr to Al–Zn–Mg–Cu alloy.[3] Sc and Zr created the complex compound $Al_3(ScZr)$. In 2013, Jin *et al.* also published their research on Al–Mg alloys with rare earth (RE) modified.[4] The composition of Vietnam's rare earth element (RE-VN) consists of two main components: La and Ce. These elements can create $Al_{11}Re_3$ and Al_2RE inter-metallic phases, $AlRE_3$ has a cube structure with a high number of sliding systems with great ductility.[4–10]

In this paper, we present the results of research on the effects of RE-VN, modification processes on the structure and hardness of Al–5Zn–3.5Mg–1.2Cu alloys.

2. Materials and Method

The materials were prepared and melted in the induction furnace with the chemical composition, as shown in Table 1. Before melting, the materials dried at 200°C to remove water. Additional elements such as Zn and Mg were calculated and prepared to carry out additions during the melting process. After the alloy was smelted and added, it was modified by 4% and 6% content of RE at 750°C for 150, 200 and 250 s. The rare-earth is introduced by submerging in liquid metal and stirring gently. The alloy is then poured into a metal mold.

After casting, the microstructure and phase transformation in samples were analyzed. Samples were tested by OM (Axiovert 25A), X-ray (Smart lab), SEM and EDS (JEOL JSM-7600F). The hardness of samples was also tested (MITUTOYO ATK-600).

Table 1. The chemical of Al–5Zn–3.5Mg–1.2Cu alloy.

Alloys	Si	Fe	Cu	Mn	Mg	Zn	La	Ce	Al
%	0.49	0.29	1.69	0.13	2.42	4.87	x	x	bal.

3. Results and Discussions

3.1. *Optical microstructure*

In Fig. 1, a comparison of the microstructures of these alloy shows that the particle size of the alloy is about 50–100 μm and the particles which are unevenly distributed can be the $MgZn_2$ or $Mg_3Zn_3Al_2$ phase. After the modification processing, there is a small branch structure that has grain size which was smaller than nonmodified. However, increasing the modification time, the particle size was increased. By OM analysis, the morphology of the phases is the common branch shape with a yellow matrix as the α-Al phase and the black part can be predicted as the inter-metallic phases of Zn, Mg, Cu along with the presence of donated soil element inside the interstitial portion of the black part. In the structure of the matrix, many tiny black dots appear that form $Al_{11}RE_3$ and Al_2RE phases.

When the content of modification changes from 4% to 6%, the particle size increases. This proves that the limit of modification content to fine-tune this alloy particle is 4%.

The X-ray diffraction (XRD) diagram in Fig. 2 shows that by adding 4% modification agents to alloy at temperature 750°C in 150 s, RE element has formed Al_2Ce phase, which is consistent with previous predictions about the phase formation structure of this super-flexible aluminum alloy system.

Based on the results of previous studies from Refs. 5 and 6, it can be seen that RE elements are concentrated mainly on the grain boundaries. Thus, the intermetallic phases with RE Al_x(La, Ce) will exist much on the grain boundary, these phases act as a barrier to limit the large particles from occurring.[6]

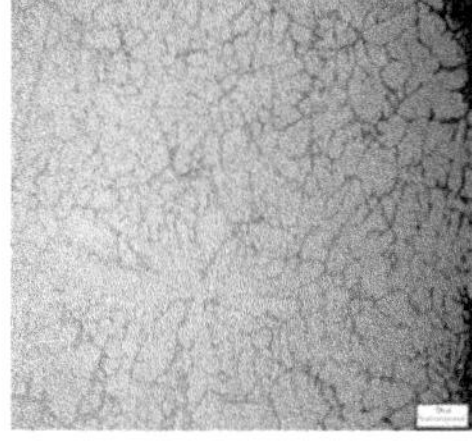
(a) Non-modification alloys
(*unit bar indicates 100 μm*)

(b) Modification alloys with
4%RE. The grain size is 6,67 μm
(*unit bar indicates 100 μm*)

(c) Modification alloys with
6%RE. The grain size is 7,97 μm
(*unit bar indicates 100 μm*)

Fig. 1.   (Color online) The microstructure of casting alloy.

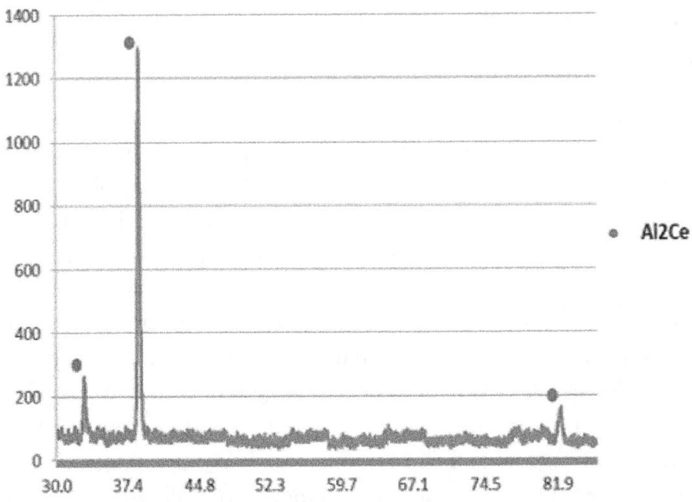

Fig. 2.   (Color online) X-ray of the sample with 4% RE.

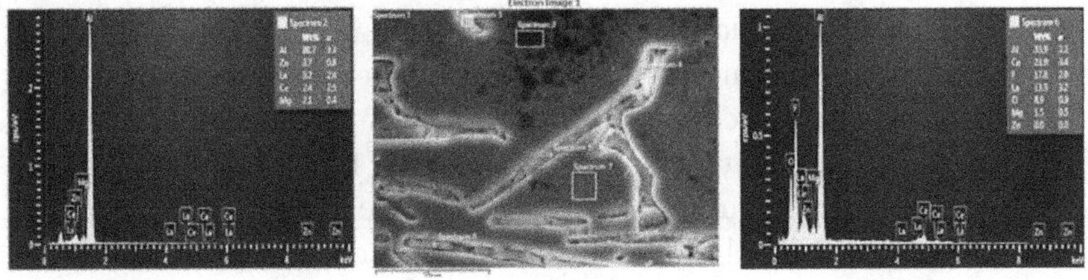

Fig. 3.   (Color online) SEM of sample 4% RE, at the 750°C and 150 s.

Figure 3 showed the SEM and EDS analysis of 4% RE modified samples, it is possible to note the presence of RE on the entire surface of the phases, both on the matrix phase as well as phase boundaries. The modification agent here is RE that does not dissolve into the liquid metal (the melting point of La, Ce is higher than the modification temperature) which is related to the reduction of energy forming crystalline sprouts, the particles of insoluble modification agents act as available phase boundaries. At location 2 in Fig. 3 and point 6 in Fig. 3, the RE is most concentrated. The modification process has also rapidly cooled, increasing the cooling process modification to achieve a finer structure, which has reduced surface tension.

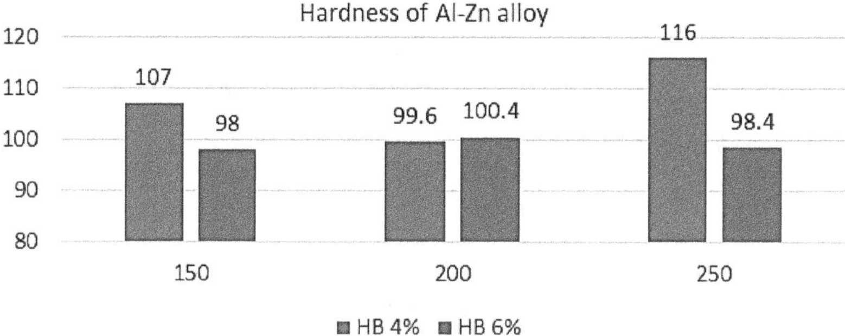

Fig. 4. (Color online) The hardness of alloy after modification.

3.2. *Hardness*

Figure 4 shows that when increasing the denaturation time, the hardness value of samples was the highest when modified at 250 s. For 6% RE modified samples, the low hardness value is explained by a large amount of modification so there may not be enough modification time to improve the properties of alloys which are explained in the microstructure of alloys. When changing the modification time for 4% RE samples due to the increased particle size and having the time to release the intermetallic phases as $MgZn_2$ or $Mg_3Zn_3Al_2$ combined with the formation of RE intermetallic phases with aluminum, the hardness value increases. However, the microstructure analysis showed that grain size of this alloy (6% RE modified sample) increased. For alloys that continue to be used for the study of superplastic alloys, the formation of small grain sizes from the casting is necessary. It showed the difference in hardness between the two samples with difference in modified content at 4% and 6%, respectively. Hardness of 4% RE modified sample is higher than the sample with a modified content of 6%.

4. Conclusion

The grain size of alloys that used modification elements decrease compared to the nonmodification. In the SEM image, there are inter-metallic phases of RE with metal which is uniformly dispersed in the metal matrix. After the modification process, the Al_2Ce phase was formed on the grain boundary, which reduces the grain size when crystallizing.

Acknowledgment

The authors would like to thank the Vietnam Maritime University and School of Material Science and Engineering, Hanoi University of Science and Technology for their support.

References

1. G. H. Strijbos and W. H. Kool, *Mater. Sci. Eng. A* **194**, 129 (1994).
2. Y. Huang *et al.*, *Mater. Sci. Eng. A* **266**, 295 (1998).
3. S. Tian *et al.*, *J. Mater. Res. Technol.* **8**, 4130 (2019).
4. L. Jin *et al.*, *J. Chem. Thermodyn.* **58**, 166 (2013).
5. A. Smolej, M. Gnamus and E. Slacek, *J. Mater. Process. Technol.* **168**, 397 (2001).
6. Y.-D. He, X.-M. Zhang, J.-H. You, *Chin. J. Nonferrous Metals* **16**, 639 (2006).
7. M. Hosseinifar, *Physical Metallurgy and Thermodynamics of Aluminum Alloys Containing Cerium and Lathanum*, Ph.D. thesis (McMaster University (2009).
8. W. Fang *et al.*, *Chin. J. Aeronaut.* **21**, 656 (2008).
9. A. Smolej, M. Gnamus and E. Slacek, *J. Mater. Process. Technol.* **118**, 397 (2001).
10. T. K. Ha *et al.*, *Mater. Sci. Eng. A* **271**, 160 (1999).

Visible-light photocatalytic activity of Fe@TiO$_2$ core–shell composite synthesized by sol–gel method[*]

Min Yen Yeh[†,§], Tzu Yuan Yang[†,¶], Tsung Chi Wu[†,‖], Shiow Yueh Lee[‡,**]
and Shun Hsyung Chang[†,††]

[†] *Department of Microelectronics Engineering,*
National Kaohsiung University of Science and Technology,
Kaohsiung 81157, Taiwan

[‡] *Department of Materials and Optoelectronic Science,*
National Sun Yat-sen University, Kaohsiung 80424, Taiwan
[§] *minyen@nkust.edu.tw*
[¶] *1061544105@nkust.edu.tw*
[‖] *F107187105@nkust.edu.tw*
[**] *sylee@mail.nsysu.edu.tw*
[††] *shchang@nkust.edu.tw*

Core–shell structure Fe@TiO$_2$ was synthesized by sol–gel method. The photocatalytic degradation of methylene blue over the Fe@TiO$_2$ reached 98% under UV light irradiation within 5 h. The band gap of the core–shell Fe@TiO$_2$ was found to have a redshift through a sintering treatment. The redshifted Fe@TiO$_2$ had a good performance of methylene-blue degradation (reaching 85%) under visible light irradiation for 5 h.

Keywords: Core/shell; sol–gel; band gap redshift.

1. Introduction

In recent years TiO$_2$ photocatalysts have received extensive attention on pollution degradation or water splitting due to their good optical and electronic properties, low cost and nontoxicity. Most of the photocatalysts including TiO$_2$ have high activity only under ultraviolet (UV) excitation.[1,2] A lower photon conversion efficiency of TiO$_2$ under solar radiation is still an issue that needs resolution. Continuous efforts have been made

---

[§]Corresponding author.
[*]To cite this article, please refer to its earlier version published in the *International Journal of Modern Physics B*, Volume 34, 2040127 (2020), DOI: 10.1142/S021797922040127X.

to improve the photocatalytic activity of TiO_2 and enhance its visible-light response by means of (1) doping them with metals such as Fe, Sn, Ag, Au, Pt and V[3,4] or (2) constructing core/shell structure of metal/TiO_2.[5,6]

In this paper, core–shall structure Fe@TiO_2 was prepared by sol–gel method. The as-prepared samples were sintered at three different temperatures of 500, 600 and 700°C for 3 h. The prepared samples have been characterized by X-ray diffraction (XRD). The microstructure of the composite materials was investigated using field emission scanning electron microscopy (FE-SEM). Absorbance analysis of the samples was conducted using UV–Vis diffuse reflectance spectroscopy (UV–Vis DRS) in the wavelength range of 200–700 nm. Methylene blue (MB) dye was chosen to be the target for photocatalytic activity analysis. The photocatalytic degradation was carried out using UV and visible-light LED irradiation sources.

2. Experimental

The Fe@TiO_2 core–shell nanoparticles were synthesized using sol–gel method. In the first step, ferromagnetic α-Fe particles were prepared by reduction of Fe^{2+} by sodium borohydride. $FeCl_2$ (0.04 g) were mixed with deionized (DI) water (50 ml) in a beaker, and then $NaBH_4$ (0.04 g in 1 ml DI water) was added dropwise to the solution. During the adding of $NaBH_4$, the solution was under ultrasonic treatment for thorough distribution, and the color of the solution turned from colorless into iron brown as Fe nanoparticles were generated. The Fe suspensor solution was poured into another beaker contained with 50 ml IPA. The beaker was then placed into a water bath at 70°C. Subsequently, $Ti(C_4H_9O)_4 \cdot$ (3.5 ml) was slowly added dropwise to the solution under vigorous stirring for 30 min to achieve Fe@TiO_2 core–shell composite. The as-synthesized Fe@TiO_2 powder was separated and collected from the suspension by centrifugation and washing by DI water for three times, and then was dried in a hot-air circular oven at 70°C for 8 h. A part of the Fe@TiO_2 was sintered in a conventional furnace at 500–700°C for 3 h, and then cooled naturally to room-temperature.

3. Results and Discussion

Figure 1 shows XRD patterns of the sol–gel synthesized Fe@TiO_2 and the samples treated with further sintering process. It can be seen that the Fe@TiO_2 sample without sintering treatment reveals amorphous structures. The XRD patterns of the Fe@TiO_2 sample sintered at 500°C showed four sharp peaks at 2 theta angle with the peak position at 25.2°, 37.7°, 47.8°

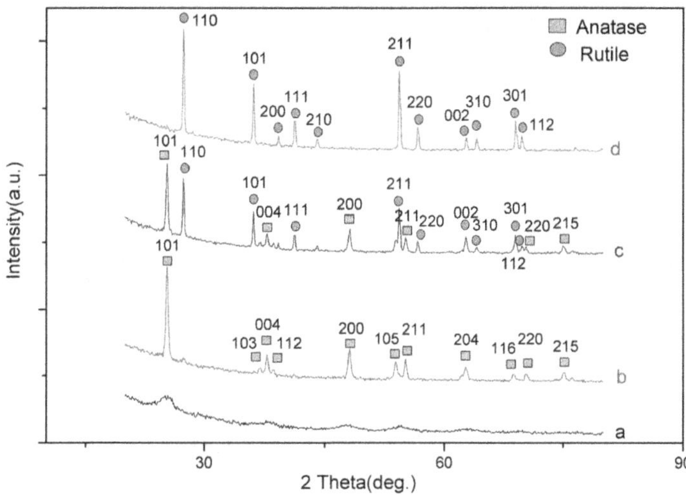

Fig. 1. (Color online) XRD pattern of Fe@TiO$_2$ (a) unsintered, and sintered at (b) 500°C, (c) 600°C and (d) 700°C.

and 54.1° with (101), (004), (200) and (105) diffraction planes, respectively. The observed four peaks match well with the tetragonal anatase of TiO$_2$ (JCPDS card no. 21-1272). For the Fe@TiO$_2$ sintered at 600°C, it could be seen that phase transition from anatase to rutile occurred, and the phase coexistence was clearly observed. As the Fe@TiO$_2$ was sintered at a higher temperature of 700°C, a complete phase conversion was obtained. The resulting structure reveals the preferred 2 theta peaks locating at 27.5°, 36.1° and 54.4° respond to the (110), (101), (211) planes of the rutile phase (JCPDS 21-1276), respectively.

Figure 2 shows the SEM micrographs of the Fe@TiO$_2$ powders. The Fe@TiO$_2$ nanoparticles were homogeneously distributed without large agglomeration. They were elliptical in shape with a diameter of 444–721 nm, as shown in Figs. 2(a)–2(c). It can be seen that the nanoparticle sizes did not exhibit obvious changes between unsintered and sintered samples. It is also worth noting that no obvious more or larger agglomerates were created through high-temperature sintering processes.

Figure 3 shows photocatalytic activity of Fe@TiO$_2$ on the degradation of MB under UV or white LED irradiation. Under UV light irradiation, the MB degradation by unsintered Fe@TiO$_2$ could reach 98% within 5 h as shown in Fig. 3(a). Figure 3(b) shows visible light photocatalytic activity and it is clear that sintered Fe@TiO$_2$ samples have significant performance

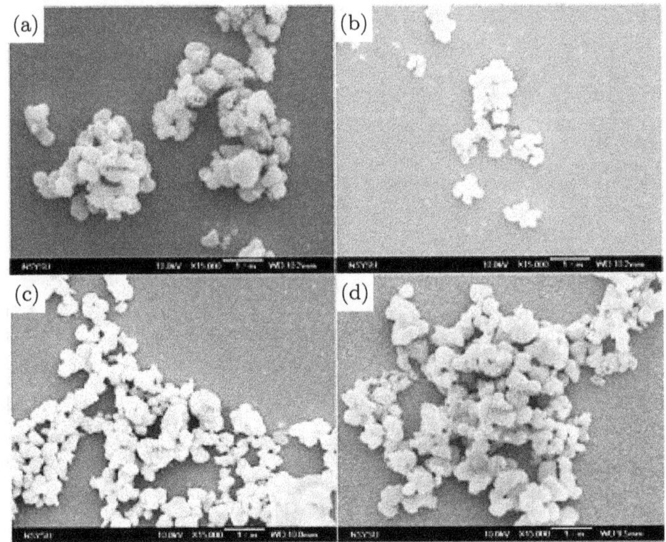

Fig. 2.   (Color online) SEM micrographs of Fe@TiO$_2$ (a) unsintered and sintered at (b) 500°C, (c) 600°C and (d) 700°C.

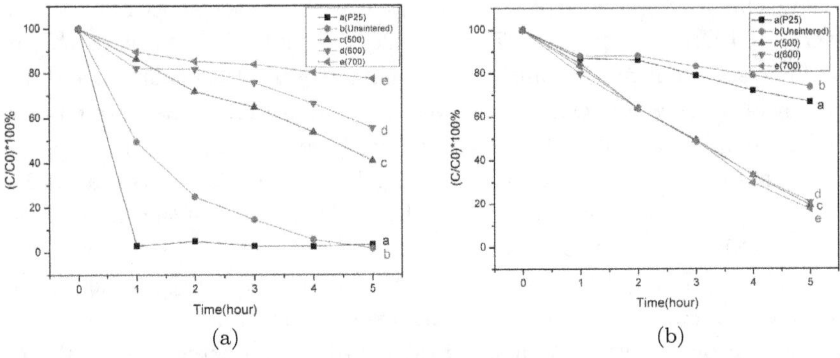

Fig. 3.   (Color online) Photocatalytic activity of Fe@TiO$_2$ on the degradation of MB under (a) UV and (b) white light LED irradiation.

reaching 85% after 5 h irradiation as compared to its UV light photocatalytic activity. From Fig. 3(b), it was also observed that the unsintered Fe@TiO$_2$ had less visible-light photocatalytic activity. This means that the unsintered sample does not like the sintered samples, which have an obvious redshift of band-gap, though it has the same core–shell Fe@/TiO$_2$ structures with the sintered samples. The only difference between the unsintered

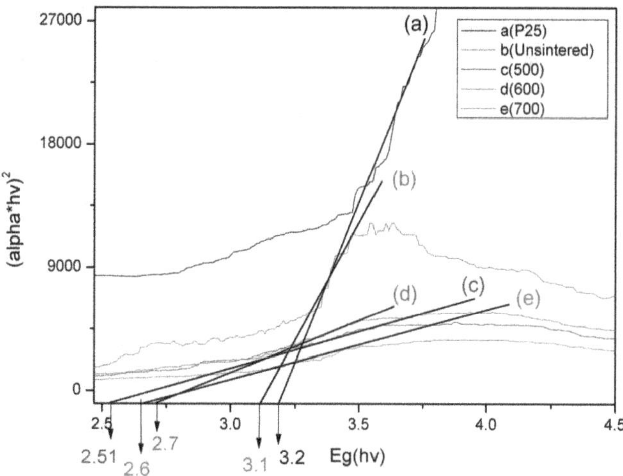

Fig. 4.   (Color online) Tauc plots of $(\alpha h\nu)^2$ versus photo energy $(h\nu)$ of (a) P25 TiO$_2$, (b) unsintered Fe@TiO$_2$ and sintered Fe@TiO$_2$ at (c) 500°C, (d) 600°C and (e) 700°C.

and sintered Fe@/TiO$_2$ is the amorphous and crystallized structures of each other.

Figure 4 shows the plot of $(\alpha h\nu)^2$ versus photo energy $(h\nu)$. The plot was derived from UV–Vis diffuse reflectance spectra. Optical E_g can be determined by Tauc equation, $(\alpha h\nu)^n = A(h\nu - E_g)$ where α is the absorption coefficient, $h\nu$ is the photon energy, A is a constant characteristic of material and n equals either $1/2$ for indirect transition and 2 for a direct transition. The E_y can be found by extrapolating a linear part of the $(\alpha h\nu)^2$ graph to the x-axis $(\alpha = 0)$ to be 3.2, 3.1, 2.7, 2.51 and 2.6 eV, respectively, for P25 TiO$_2$, unsintered Fe@TiO$_2$, and sintered Fe@TiO$_2$ at 600, 500 and 700°C. It is clear that the redshift in band gap obviously only happens to the sintered core–shell Fe@TiO$_2$ samples with crystallized structures.

4. Conclusion

Fe@TiO$_2$ core–shell nanoparticles were successfully synthesized by sol–gel method. The samples were well crystallized after sintering and their phase was strongly dependent on the sintering temperature. The sintered core–shell Fe@TiO$_2$ with a redshift of band gap exhibits a significant photocatalytic activity under visible light irradiation. Amorphous Fe@TiO$_2$ did not have an obvious redshift of bandgap though it had core–shell structures.

References

1. V. Vaiano, O. Sacco and M. Matarangolo, *Catal. Today* **315**, 230 (2018).
2. M. R. Al-Mamun *et al.*, *Environ. Chem. Eng.* **7**, 103248 (2019).
3. N. Sobana, M. Muruganadham and M. Swaminathan, *Mol. Catal. A Chem.* **258**, 124 (2006).
4. J. Zhu *et al.*, *Mol. Catal. A Chem.* **216**, 35 (2004).
5. G. K. Naik *et al.*, *J. Alloys Compd.* **771**, 505 (2019).
6. H. Fu *et al.*, *Powder Technol.* **328**, 389 (2018).

A research on relevance between photoplethysmography signal and perceptual stimulation*

Chih-Hsueh Lin, Zhi-Hao Wang[†] and Gwo-Jia Jong

Department of Electronic Engineering,
National Kaohsiung University of Science and Technology,
Kaohsiung 80778, Taiwan, ROC
[†] *zhwang0401@hotmail.com*

In this paper, the Photoplethysmography (PPG) signal response of humans under different perceptual stimuli was mainly discussed. The autonomic nervous system (ANS) regulates the functioning of organs or tissues in the body to make the body adapt to the environment. When the human body is stimulated, it can obtain information about ANS from the analysis results of heart rate variability (HRV). The proposed method is used with PPG signals for time domain and frequency domain analysis to obtain HRV parameters and short-time Fourier transform (STFT) output heat map, respectively. Different environmental stimuli were given during the experiment and changes in PPG signals were observed. There were eight males and eight females with a total of 16 subjects with an average age of 24 years. Heart rate was observed to vary with stimulation in 16 subjects. The heat map clearly distinguishes whether the subject is affected by the stimulus and causes the ANS to regulate the human body. The data obtained under various stimuli are directly compared and measured from physiological signals. It can establish an objective patient's current emotional judgment and can be used to explore the patient's diagnosis status during the diagnosis process.

Keywords: Photoplethysmography (PPG); autonomic nervous system (ANS); the short-time Fourier transform (STFT); perceptual stimulation.

[†]Corresponding author.
*To cite this article, please refer to its earlier version published in the *International Journal of Modern Physics B*, Volume 34, 2040129 (2020), DOI: 10.1142/S0217979220401293.

1. Introduction

In the human body, the autonomic nervous system (ANS) controls the neurons of many organs to maintain the balance of the body.[1,2] The heart motion is also controlled by the ANS.[3] In this paper, the mental state of the subject was evaluated by additional stimulation and multi-cycle observation of cardiac activity. Cardiac activity monitoring is monitored by a pulse oximeter measuring Photoplethysmography (PPG) signals.[4] The aim of this paper is the analysis of PPG signals from our proposed wireless transmission architecture that consists of a pulse oximeter and a wireless transmitter.[5] The normal PPG signal is composed of two waveforms, which are waveforms caused by systolic and diastolic pressures.[6] The concentration of oxygenated and deoxygenated blood is controlled by the contraction and relaxation of the heart.[7] Therefore, measuring the oxygen concentration in the blood by a pulse oximeter can measure two peaks in a complete cycle. The measured PPG signal is transformed by fast Fourier transform to obtain the spectrum.[8]

Optical sensors on PPG signals include light sources, usually light-emitting diodes and photodetector elements.[9,10] The light source is used to illuminate the tissue area, and the photodetector measures the amount of light leaving the tissue at different positions. The PPG signal consists of systolic blood pressure and diastolic blood pressure.[11] When the pressure in the heart constricts the blood vessels, the volume of blood in the formation period will change continuously. If the diastolic pressure of the heart is relatively small, the output of the previous systolic blood circulation produces the effect of the heart valve reflex phenomenon.

2. Materials and Methods

The block diagram of the research methodology is shown in Fig. 1. In general, the sensitivity of the ANS is related to age. In order to reduce the factors in the experimental process, the participants are eight males and eight females, with an average of 24 years. All the participants were from the same laboratory, which ensures the same lifestyle background and fairness of the results. The PPG measurement module is placed on the finger of the right

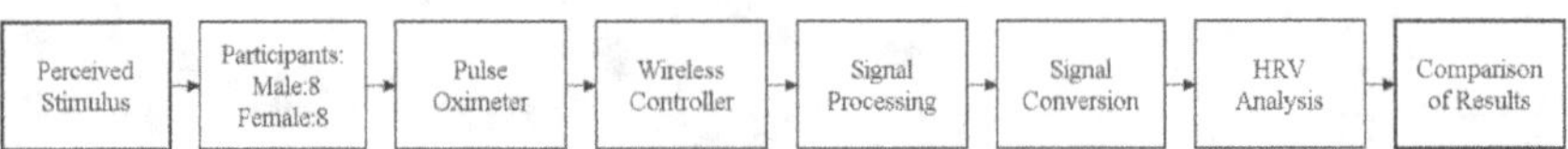

Fig. 1.   The block diagram of research methodology.

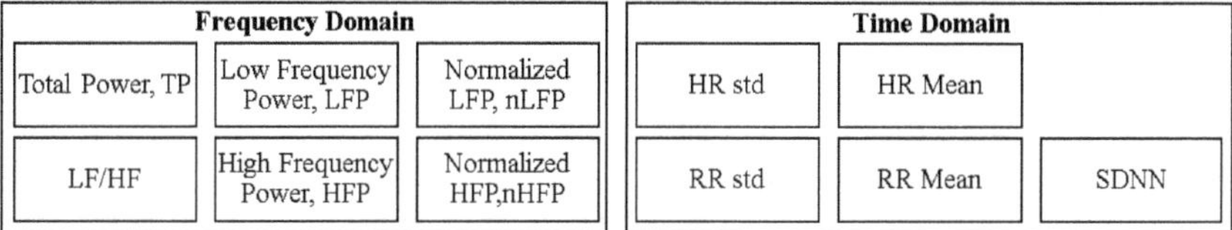

Fig. 2. The parameter diagram of HRV.

hand which consists of a pulse oximeter and a controller (ESP8266) with a Wi-Fi protocol. The output of the pulse oximeter is a 0–5 V analog voltage signal that reflects changes in blood oxygen concentration. This analog signal is converted into a digital signal by the controller's A/D converter. The measured PPG signal is transmitted to the server for synchronization and signal processing.

On the server, heart rate variability (HRV) parameters are obtained from time-domain analysis and frequency-domain analysis, as shown in Fig. 2. The time-domain analysis is a numerical analysis of heart rate (HR) and R–R interval (RR) which are based on statistics, which mainly calculates standard deviation, mean value or number of variations. A typical frequency domain analysis of HRV uses Fourier transform (FT) or fast Fourier transform (FFT) to convert time-series data to spectrum distribution. The common parameters are a power of Very low frequency (VLF), Low frequency (LF), High frequency (HF) as well as the ratio of LF and HF. As the PPG signal is a periodic signal and each may contain different information, the short-time Fourier transform (STFT) analyzes signals of different segments at the same time that is more suitable. The mathematical expression of STFT is shown in Eq. (1), where $x(t)$ is input time-series signal, ω is the angular frequency, $w(t - \tau)$ is a window function, and τ is time.

$$\text{STFT}\{x(t)\}(\tau,\omega) \equiv X(\tau,\omega) = \int_{-\infty}^{\infty} x(t)w(t - \tau)e^{-j\omega t}dt. \tag{1}$$

The experiment was conducted at 10 am for 12 consecutive days, and the physical condition of this period was quite good for a normal worker. Our test is divided into three modes. Each mode will be tested for four days, and the test mode of the previous day and the next day will not be the same. Each test time is 10 min in total. The first 5 min measure the PPG reference signal without any stimulation, and the last 5 min measure the PPG signal with the perceived stimulus. After the complete experiment, one subject will have 12 general PPG signals and 12 stimulating PPG signals.

3. Results

The PPG signal represents a record of cardiac activity. In this study, experiment groups were divided into two groups, namely the control group and the study group. Participants were not given stimulation in the control group but were given stimulation in the study group. Figure 3 shows the heat map of the PPG signal after STFT conversion. From the heat map, we can observe the relationship between the X-axis (time), Y-axis (frequency), and color (amplitude). From the X-axis, it can be observed that within 5 s, the control group and the experimental group recorded 7 and 9 PPG waveforms, respectively. The frequency information implied by each PPG waveform can be observed on the Y-axis. Different colors indicate the intensity of the conversion result, the results can be observed from the waveform that each pattern feature contains a red or yellow feature (primary frequency).

Since PPG signals and ECG signals could be cross-referenced in past studies, HRV was used to analyze PPG signals in this research. Table 1 shows the analysis results of PPG signals, which are divided into time-domain analysis and frequency domain analysis.

The time-domain mainly shows the beating variation of the heart. The results observed show that the mean HR increases and the RR interval

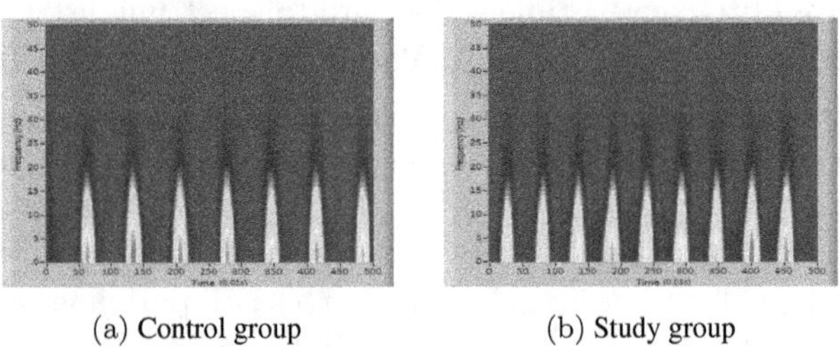

(a) Control group   (b) Study group

Fig. 3. (Color online) The heat map of PPG signal after the STFT process.

Table 1. Measurement of physiological parameters for one of the participants.

	Time-domain			Frequency-domain		
Parameter	Control group	Study group	Parameter	Control group	Study group
SDNN	30.44	42.12	VLF	21%	16%
HR std	4.347	36.83	LF	48%	42%
HR mean	87.09	125.1	HF	31%	42%
RR std	0.02843	0.07614	LF/HF	1.6	1
RR mean	0.6904	0.5028			

decreases after the study group receives stimulation as well as the standard deviation of the heart rhythm also increases. This represents arrhythmia and irregular heart cycle intervals when participants are stimulated. The frequency-domain shows the intensity of autonomic nerve activity. From the results, it can be observed that the regulation of sympathetic nerves of the participants is weakened, in other words, the activity of sympathetic nerves decreases, and the activity of parasympathetic nerves increases. On the whole, it can be observed from Fig. 3 and Table 1 that when participants receive stimuli, they will cause a rapid heartbeat, which will affect the activity of the autonomic nerve. This series of physiological activities are presented quantitatively and graphically in this research.

4. Discussion

The results show that the heart rate of 16 participants increased significantly after being stimulated, and the other HRV parameters were also significantly affected. In the 12-round test, two participants were significantly affected eight times, eight participants had 11 significant effects, and six participants had 12 significant effects. Therefore, it can be observed from the experimental results that the information obtained after the analysis of the PPG signal can determine whether the participants have been stimulated or interfered with by the environment. ANS controls many physiological responses, including heartbeat. When the participant feels stress or sensory stimuli, the sympathetic nerves in the ANS stimulate heart rate increases, while breathing rhythm and muscles are adaptively adjusted.

5. Conclusions

In general, cardiac motion is controlled by the sympathetic nerves in the ANS. When the human body is exposed to environmental changes, ANS regulates the functioning of organs or tissues in the body to make the body adapt to the environment. When the human body is stimulated, it can obtain information about ANS from the analysis results of HRV. A method was proposed for obtaining HRV parameters and heat maps using PPG signals for time domain and frequency domain analysis, respectively. At the same time, the process of HRV is visualized through heat maps to enhance the effect on heart signals. Different environmental stimuli were given during the experiment and changes in PPG signals were observed. Heart rate was observed to vary with stimulation in 16 subjects. The heat map clearly distinguishes whether the subject is affected by the stimulus and causes the ANS to regulate the human body.

References

1. M. C. Tiveron, M. R. Hirsch and J. F. Brunet, *J. Neurosci.* **16**, 7949 (1996).
2. M. Cui *et al.*, *J. Neurosci.* **39**, 5816 (2019).
3. A. J. Fourcin *et al.*, *Br. J. Audiol.* **13**, 85 (1979).
4. P. M. Mohan *et al.*, Measurement of arterial oxygen saturation (SpO_2) using PPG optical sensor, in *2016 Int. Conf. Communication and Signal Processing (ICCSP)* (IEEE, 2016), pp. 1136–1140.
5. M. Elgendi, *Curr. Cardiol.* **8**, 14 (2012).
6. D. McDuff, S. Gontarek and R. W. Picard, *IEEE Trans. Biomed. Eng.* **61**, 2948 (2014).
7. R. Gordan, J. K. Gwathmey and L. H. Xie, *World J. Cardiol.* **7**, 204 (2015).
8. M. Pirhonen, M. Peltokangas and A. Vehkaoja, *Sensors* **18**, 1693 (2018).
9. J. Allen, *Physiol. Meas.* **28**, R1 (2007).
10. Y. H. Kao, C.-P. Chao and C. L. Wey, A PPG sensor for continuous cuffless blood pressure monitoring with self-adaptive signal processing, in *2017 Int. Conf. Applied System Innovation (ICASI)* (IEEE, 2017), pp. 357–360.
11. M. A. Islam and M. Ahmad, Design and implementation of non-invasive continuous blood pressure measurement and monitoring system using photoplethysmography, in *2018 10th Int. Conf. Electrical and Computer Engineering (ICECE)* (IEEE, 2018), pp. 173–176.

Comparison of water uptake behavior on tensile property of epoxy impregnated continuous basalt and slag filament composites*

Se-Yoon Kim

Major of Materials Engineering, Korea Maritime and Ocean University, Busan 49112, Republic of Korea
seyun8269@naver.com

Soo-Jeong Park

Department of Mechanical and Aerospace Engineering, Naval Postgraduate School, Monterey CA 93943, USA

Chang-Wook Park

Research Institute of Medium and Small Shipbuilding, 38-6, Noksan Industrial Complex 232-ro, Gangseo-gu, Busan 46757, Republic of Korea
pcw0591@naver.com

Yun-Hae Kim[†]

Department of Ocean Advanced Materials Convergence Engineering, Korea Maritime and Ocean University, Busan 49112, Republic of Korea
yunheak@kmou.ac.kr

Slag fiber has economic and environmental advantages in that it converts a low-value-added material to a high-value-added material. However, although the slag fiber has a chemical composition similar to basalt fiber, its competitiveness in the fiber industry is significantly lower. Moreover, the slag fiber remains in the pre-commercial stage due to the uncertainty and instability of the basic properties. Therefore, in this study, the slag fiber customized through the fiberization process was compared with the existing basalt fiber to analyze the effect of the similarity of chemical composition on the environmental

---

[†]Corresponding author.
*To cite this article, please refer to its earlier version published in the *International Journal of Modern Physics B*, Volume 34, 2040130 (2020), DOI: 10.1142/S021797922040130X.

degradation characteristics and mechanical properties under tensile loading. As a result, the slag filament composites showed lower tensile strength due to the weaker interfacial bonding strength with the epoxy matrix than the basalt filament composites, but the difference in decreased tensile strength rate was not significant. In addition, long-term moisture absorption in fresh water and seawater demonstrated excellent moisture absorption resistance.

Keywords: Polymer composites; basalt fiber; slag fiber; water absorption behavior; mechanical property.

1. Introduction

Polymer composites are a promising material for structure and are widely utilized in various industries even under extreme conditions like an acid-base, seawater, corrosion environment. With a high durability, it has complex failure behavior and design diversity of constituent materials.[1] Recently, there have been attempts to transition from a paradigm that relies on high specifications and quality in terms of economic and environmental efficiency. Despite various studies on biomaterials, there are still conflicting aspects such as lifespan and safety of the materials, and pre- and post-environmental hazards when disposing the materials depending on the type of fiber reinforcement in polymer matrix fiber reinforced plastic (FRP) composite. In this aspect, basalt fiber based on natural rock and slag fiber produced through refining and fiberization from slag, a by-product of the steelmaking process, exhibits a similar chemical composition and also has a relatively superior comparative advantage to carbon fiber and glass fiber with high compatibility.[2,3] However, the slag fiber has not reached the commercialization stage. And this slag fiber, compared with basalt fiber, still lacks a practical research such as physical performance, environmental resistance, chemical stability, and functionality for structure, and it also requires a feasibility analysis on an alternative materials to the existing fiber reinforcement.[4] Moreover, the similarity of the physical properties of basalt fiber and slag fiber is an important factor in securing the potential impact of slag fiber, and it will contribute to establishing a practical research infrastructure through material-based characteristic analysis under various loads and environmental conditions.

Therefore, in this study, to determine the effect of chemical composition on physical properties, customized slag fiber was used to analyze the similarity with basalt fiber on the environmental degradation characteristics. And it was carried out through water uptake behavior under fresh

Table 1.   Chemical composition of basalt fiber and slag fiber.

Element (wt.%)	Si	Al	Ca	Fe	Na	Mg	Ti	K	C
Basalt fiber	62.5	9.2	8.5	1.8	7.0	7.8	1.4	1.8	1.6
Slag fiber	48.3	12.5	5.9	5.1	7.7	8.9	1.8	3.1	6.7

water/seawater environment. Furthermore, the mechanical properties of the basalt fiber and slag fiber were compared under tensile load and their failure causes were investigated.

2. Experimental Works

Two types of fiber reinforcements used for epoxy impregnated filament composites were basalt fiber of GM Composite Co., and customized slag fiber, which had a composition mixed with fly ash:furnance: Fe–Ni at 50:25:25 each and radiated at 16 m/s speed at an average temperature of 1221°C through fiberization process. The chemical composition of these fibers is shown in Table 1. The polymer matrix was mixed with KFR-120V epoxy and KFH-141 hardener supplied from Kukdo Chemical Co., LTD, at a ratio of 72:28. The water uptake behavior of these basalt fiber reinforced plastic (BFRP) and slag fiber reinforced plastic (SFRP) under fresh water/seawater absorption environment was measured through Eq. (1) of Fickian Diffusion.

$$M = \frac{W_m - W_d}{W_d} \times 100 \,, \tag{1}$$

where M is the absorption rate (%), W_d is the initial specimen weight (g) and W_m is the specimen weight after immersion (g).

The tensile test was conducted with a removal-oriented tensile specimen, which had a fiber content of about 40–50 wt.%. These tests were by ASTM D 5229 and ASTM D 4018, respectively. The surface of the composites was observed through SEM (Scanning electron microscope, MIRA-3 of Tescan Co.), to investigate the effect on damage behavior during fresh water/seawater immersion.

3. Results and Discussions

3.1. *Moisture absorption behavior of BFRP and SFRP*

As shown in Fig. 1(a), BFRP and SFRP have an increased absorption rate in both fresh water and seawater, and maintained equilibrium by reaching

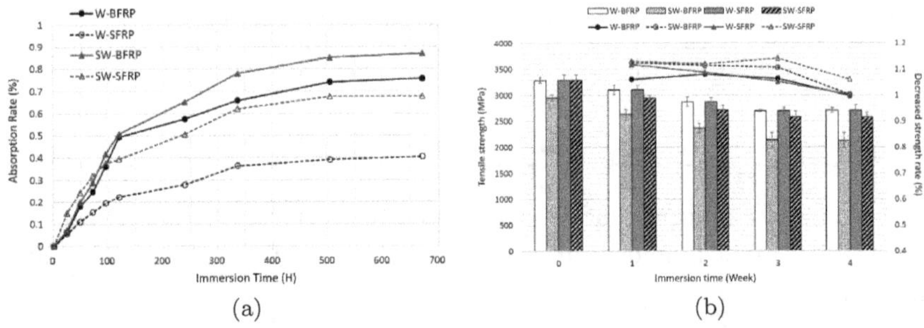

Fig. 1.   (Color online) Comparison of (a) water absorption rate and (b) tensile strength between BFRP and SFRP under fresh water/seawater environment (W: under fresh water, SW: under seawater).

water saturation state after 350 h. BFRP showed relatively high absorption rate in fresh water and seawater, whereas SFRP exhibited excellent hygroscopicity in fresh water. In terms of moisture absorption resistance, SFRP was 46.75% and 22.2% higher than BFRP in fresh water and seawater, respectively. SW-BFRP and SW-SFRP tended to be 14.9% and 67.2% lower than W-BFRP and W-SFRP. As a result, BFRP and SFRP were observed to be vulnerable in seawater. This is considered to be due to the relatively lower diffusion activity energy of seawater than that of freshwater at the same temperature.[5] This resulted in an accelerated phenomenon of interfacial erosion due to salinity in the seawater, despite the fact that seawater contained a larger amount of ions than water molecules when compared to fresh water.

3.2. *Mechanical properties and failure analysis under tensile loading*

Figure 1(b) shows the change in tensile strength over time exposed to fresh water and seawater, and BFRP and SFRP commonly displayed gradual strength deterioration in fresh water and seawater. BFRP has a relatively higher tensile strength than SFRP, and a low strength decreased rate was measured. Both BFRP and SFRP were more vulnerable to seawater than fresh water, and this is because salts and dissolved ions in seawater mainly cause polymer swelling and weaken the interfacial bonding force of the laminated composites. W-BFRP, however, maintained a low decreased strength rate despite the rapid inflow of large amounts of water based on strong interfacial bonding between epoxy and basalt fiber in neat BFRP.[6]

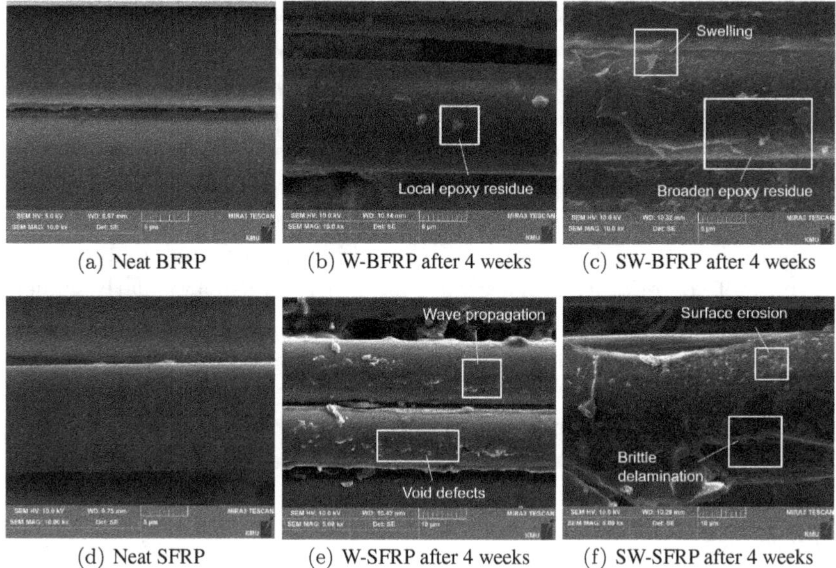

(a) Neat BFRP (b) W-BFRP after 4 weeks (c) SW-BFRP after 4 weeks

(d) Neat SFRP (e) W-SFRP after 4 weeks (f) SW-SFRP after 4 weeks

Fig. 2. Surface morphological changes through SEM observation.

3.3. *Morphological structure observation*

Figure 2 shows the fiber surface exposed at the composite interface through SEM, when the moisture absorption saturation is reached. A large amount of residues were commonly observed on the surface of SW-BFRP and SW-SFRP, and distributed along the interface with the torn epoxy. It is considered that not only moisture but also ions contained in seawater are directly related to deterioration. In particular, swelling and fatal damages such as crack growth occurred in W-SFRP and SW-SFRP, which have relatively little moisture absorption, and delamination was notable at the fiber-epoxy interface in SW-SFRP. It demonstrates lower tensile strength than BFRP.

4. Conclusions

(1) SFRP has relatively superior moisture absorption resistance compared to BFRP, and exhibits a significantly lower moisture permeability, which increases moisture absorption resistance by up to 67% in conditions where interfacial erosion is not active, such as fresh water.

(2) The damage to the fiber surface was a major factor ultimately affecting moisture absorption resistance. This was involved in promoting the acceleration of the initial moisture absorption due to the increase in the water permeation rate, but it did not affect the time taken to reach the final moisture saturation in the composites by moisture absorption.

(3) BFRP has a relatively high interfacial bonding strength compared to SFRP, which ultimately affects the low decreased tensile strength rate in fresh water and seawater. BFRP and SFRP were commonly used to promote deterioration in seawater, and SFRP was particularly vulnerable to even amounts of moisture.

(4) Customized slag fiber showed high utilization value as industrial fiber by deriving the moisture absorption resistance corresponding to basalt fiber in extreme water environment. However, optimization of the design process for fiberization of slag is required to improve durability.

References

1. R. Roy, B. K. Sarkar and N. R. Bose, *Bull. Mater. Sci.* **24**, 87 (2001).
2. Q. Liu *et al.*, *Polym. Composit.* **27**, 41 (2006).
3. S. Y. Kim, *Study on Evaluation of Mechanical and Environmental Properties of Slag Fiber Reinforced Composites* (Busan, Republic of Korea, 2019).
4. C. W. Park and Y. H. Kim, Fiberization and reuse of slag for high added value and its application, in *Int. Conf. Physics and Mechanics of New Materials and Their Applications* (Springer, Cham, Switzerland, 2017), pp. 589–603.
5. W. R. Broughton and M. J. Lodeiro, Techniques for monitoring water absorption in fibre-reinforced polymer composites. Available from http://midas.npl.co.uk/midas/content/mn064.html (National Physics Laboratory, UK).
6. S. J. Park *et al.*, *Mod. Phys. Lett. B* **33**, 1940020 (2019).

Properties of the interface between the as-built Ti–6Al–4V sample and the Ti substrate in selective laser sintering*

Minh-Thuyet Nguyen[†,§], Hoang-Viet Nguyen[†], Thai-Hung Le[†],
Quoc-Khanh Dang[†] and Jin-Chun Kim[‡]

[†]*School of Materials Science and Engineering,
Hanoi University of Science and Technology,
No. 1, Dai Co Viet Street, Hai Ba Trung District,
Hanoi 100000, Vietnam*

[‡]*School of Materials Science and Engineering,
University of Ulsan, Ulsan 44610, Korea*
[§]*thuyet.nguyenminh@hust.edu.vn*

In this paper, selective laser sintering (SLS) was applied to join two materials by printing Ti–6Al–4V powder on a Ti substrate without any support parts. The characteristics of the interface between the as-built Ti–6Al–4V sample and the substrate were investigated. The analysis indicates that a heat-affected zone (HAZ) and the fish-scale type were formed at the joining area. The combination of smaller grains, acicular α' martensite, and lamellar $(\alpha + \beta)$ structures was observed inside the interface zone. The hardness value at the interface area was measured by about 320 HV which is higher than 280 HV of the substrate and smaller than 369 HV of the as-built sample. The results predict that the SLS process is a promising method for manufacturing of hybrid materials.

Keywords: Selective laser sintering; interface; Ti–6Al–4V alloy; titanium.

1. Introduction

Selective laser sintering (SLS) is a type of additive manufacturing (AM) that has been used to produce fully dense parts and complex geometry objects from metal powder by using a laser as the power source.[1] It is also known as a method that reduces processing time in manufacturing the

---

[§]Corresponding author.
*To cite this article, please refer to its earlier version published in the *International Journal of Modern Physics B*, Volume 34, 2040137 (2020), DOI: 10.1142/S0217979220401372.

replacement parts. Therefore, SLS has become a promising technique used in various applications where the unique design is required like automotive, aerospace and medical-dental fields. Up to now, Ti–6Al–4V alloys have been widely used as the charging material in the SLS process to fabricate parts for investigating and applying.[1–3] Generally, the SLS process is operated to build Ti–6Al–4V parts upon a substrate with a support structure in order to reduce the time and the difficulty of the separation steps. However, the SLS of Ti–6Al–4V powders directly on the substrate without support parts could also be operated. This optional processing suggests a new way to join materials, repair damaged components or the broken parts in a short time by depositing materials directly on the fractured surfaces.[4–7] Once this processing is used, the formation and characteristics of the welding between the as-built sample and the substrate are very important and it must be focused on investigation. However, most of the recent studies concentrated on evaluating the microstructure and mechanical properties of the as-built Ti–6Al–4V parts, while the as-built Ti–6Al–4V/substrate behavior in SLS process is poorly understood. Therefore, in this work, the microstructure and mechanical properties of the interface between the as-built Ti–6Al–4V sample and the Ti substrate which was formed in SLS of atomized Ti–6Al–4V powders directly on the Ti substrate without support structures were fully investigated.

2. Materials and Methods

Gas-atomized Ti–6Al–4V powder with a particle size of under 45 μm was used as the feeding material in SLS process. All the samples in this work were built directly on a Ti substrate without any support parts to form the cubic shapes of $10 \times 10 \times 10$ mm. The parameter conditions are presented in Table 1. Layers were scanned using the continuous laser mode in a zig–zag pattern.

For metallographic analysis, the specimens were roughly polished using 300, 400, 600, 800 and 1200 grits of silicon carbide (SiC) papers, followed by the final polishing with alumina suspensions (3, 1 and 0.05 μm).

Table 1.   Parameter conditions of SLS process used in this work.

Power (W)	Scan speed (mm/s)	Laser beam diameter (μm)	Nitrogen %	Hatch distance (μm)	Layer thickness (μm)
120	600, 800, 1000	90	99.9	70	80

Optical microscopy (OM-Olympus PMG3) and field emission scanning electron microscopy (FE-SEM — JEOL JSM-6500F) were used to observe the surface morphology of the specimens. Microhardness test was done by using the micro-Vickers hardness (Mitutoyo MVK-H1) with a load of 500 g.

3. Results and Discussions

In this work, samples were cut out with the remaining substrate part for each one after the SLS process. The microstructural morphology was observed at the body part, the substrate part and the joint part between the as-built part and the substrate.

In general, the microstructure of the as-built body part is observed at the top-view and the side-view surfaces. As we have reported in our previous study,[8] the top-view surface shows bi-directional scan tracks, molten pool boundaries and the microstructure features include grains and α' phases in each molten pool. The side-view reveals the microstructure consisting of a martensitic α'-phase within long, columnar grains which are identified as prior β-grains that grow during the solidification and orient more or less in the building direction, which is the typical microstructure of the as-built Ti–6Al–4V products.[9,10] In this work, the welding of the printed part on the substrate was investigated, therefore, the cross-section of the sample without the separation of the substrate part was taken out for the evaluation process.

Figure 1 shows a good welding surface of the bonding areas between the as-built samples and the substrate. Thus, these areas exhibit a homogenous matrix without any separated borders, which demonstrates that powders were sintered and welded well on the substrate.

The change in microstructures along the build direction from the substrate to the as-built body part was observed, as shown in Figs. 2–4. From these images, we could observe three areas including the substrate, the

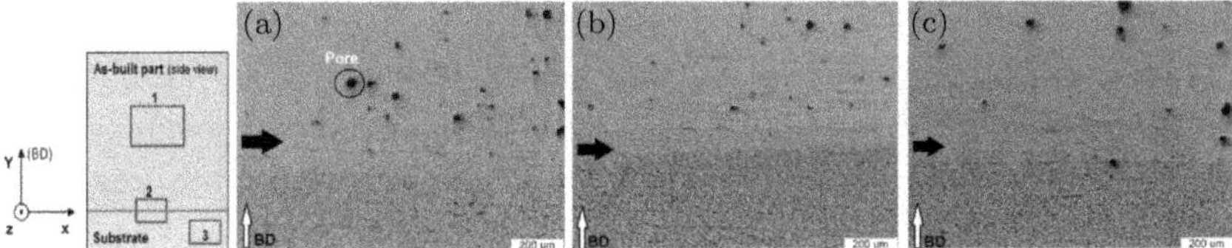

Fig. 1. OM images show the welding areas (black arrow) between as-built (upper part) and substrate (lower part) in different power-speed parameter conditions: (a) 120 W–1000 m/s; (b) 120 W–800 m/s and (c) 120 W–600 m/s according to the order in Table 1.

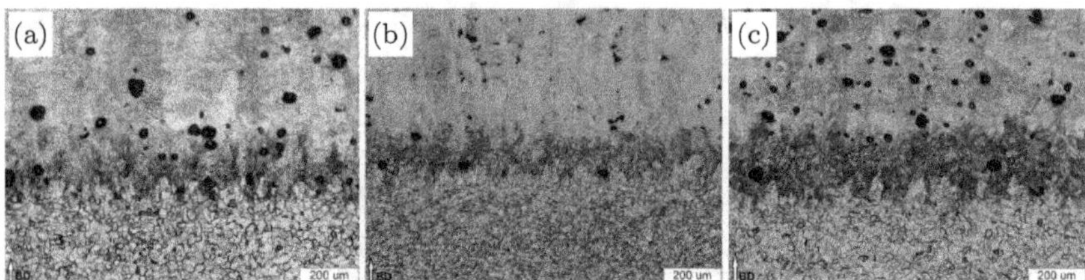

Fig. 2. OM images show the interface area between the as-built sample and Ti substrate.

Fig. 3. FE-SEM images show: (a) the fish-scale feature, (b) HAZ region and (c) interface area.

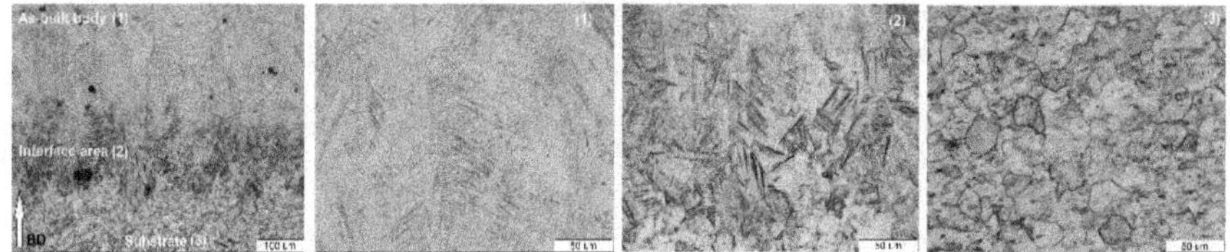

Fig. 4. OM images show the change in microstructure from the substrate to the as-built part.

interface and the as-built part areas (Figs. 2 and 3). The interface area consists of a heat-affect zone (HAZ) and a fish-scale structure region (Fig. 3). The fish-scale structures can be observed in other materials fabricated by AM such as 316L stainless steels, AlSi, and Inconel 625 alloys but not normally observed in the as-built Ti–6Al–4V body sample.[2,12] Therefore, from our results, it could be concluded that the fish-scale structure was only formed due to the thermal diffusion from the beginning of scanning layers right above the substrate. Figure 4 shows that the microstructure of the substrate comprised of primary α grains within a small lamellar $(\alpha + \beta)$, Fig. 4(3) and the as-built sample expose the columnar prior-β grains with acicular α' martensite inside, Fig. 4(1). The interface area shows a different microstructure with the presence of primary α, acicular α' martensite, lamellar $(\alpha + \beta)$, and partially decomposed α' phases, Fig. 4(2). The formation of these structures at the interface area is attributed to the interaction between the moving laser beam, the powders and the substrate that leads to complex physical phenomena such as absorption and reflection of the laser beam,[13] melting, solidification, re-melting, and re-solidification.[2,3,14] Thus, melting is the first phase-transformation process involved in SLS, and the maximum metal pool temperature during AM was about $2000°C^{11}$ while the liquidus temperature of Ti–6Al–4V was $\sim 1650 \pm 15°C$ and its solidus temperature was $\sim 1605 \pm 10°C$.[15] Therefore, the in SLS process, when laser beams scanned on the first layers of powders spread on the substrate surface, powders and a small amount of the substrate material melted together, leading to the welding of the sintered layer on the substrate and the formation of the HAZ zone. After the melt pool was formed, the cooling and solidification took place. It is noted that the maximum cooling rate of Ti–6Al–4V in the melt pool is in the range of $(1.2–4.0) \times 10^4 \; °C/s^{2,11}$ and the initial cooling rates in the as-built sample near the melt pool were found to be extremely higher than the critical cooling rate $(>18–23°C$ or $>410°C)^{16–18}$ that required martensitic transformation in Ti–6Al–4V. Due to that rapid cooling regime, α' martensite is thus commonly formed and observed in the as-built sample as well as in the built sample/substrate interface area. As could be observed in Fig. 4(2), the existence of the martensitic microstructure was dominant at the interface area, however, it can decompose into partially decomposed α' and fully lamellar $(\alpha + \beta)$.[2,19] Also, transformed beta grains with acicular alpha grains can be formed in the joined area due to the increase of heat input from the laser beam (Fig. 4).

Besides the observation of the microstructure, the mechanical property was also investigated. Figure 5 illustrates the distribution of hardness along the build direction from the substrate to the as-built body part. We found that the average hardness of the weld zone was ~ 320 HV which is higher than that for the substrate material $(\sim 280$ HV) but smaller than that for the as-built body part $(\sim 369$ HV).

The existence of the primary martensitic α' phase is the most important factor for the high value of hardness in the as-built body sample.[2,16,19] The increase in hardness of the weld area compared to the substrate was reported in the formation of fine grains and harder phases of α' and lamellar $(\alpha + \beta)^{19,20}$ in the interface region while the substrate material had the structure of the equiaxed grain structures with primary α and β.

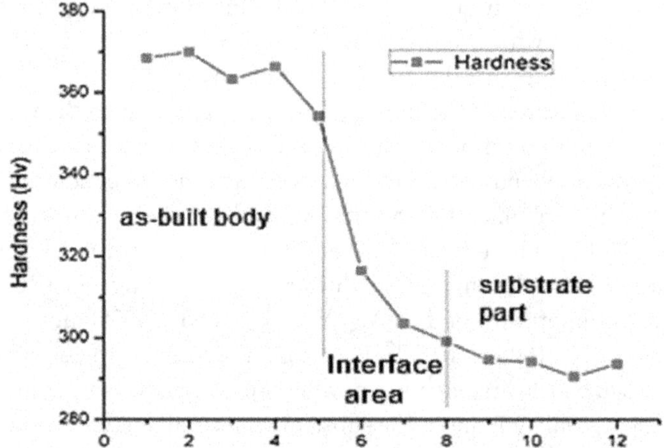

Fig. 5.   Microhardness distribution in the sample along building direction.

In addition, according to the Hall–Petch effects,[19,20] the combination of the fine grains, acicular α' martensite, lamellar $(\alpha + \beta)$ phases could help in strengthening the joint part between the as-built sample and the substrate.

4. Conclusions

A comparative study was conducted to understand the characteristics of the interface between the as-built Ti–6Al–4V sample and the substrate in the SLS process without support parts. Based on the investigations performed, the following conclusions can be drawn:

— The as-built Ti–6Al–4V sample was built and welded strongly on the Ti substrate material due to an interface zone within the fish-scale and HAZ areas.

— The microstructure of the substrate materials comprised primarily of equiaxed α grains, while the interface zone displayed a combination of smaller grains, acicular α' martensite, and lamellar $(\alpha + \beta)$ phases.

— The hardness value of the interface was higher than that of the substrate but smaller than the hardness of the as-built sample.

— The interfacial area connected the substrate to sintered body perfectly without any separation, which predicted the potential of using this technique for joining and repairing parts.

Acknowledgment

This research was funded by Vietnam National Foundation for Science and Technology Development (NAFOSTED) under Grant Number 107.99-2018.336.

References

1. C. Y. Yap *et al.*, *Appl. Phys. Rev.* **2**, 041101 (2015).
2. M. Qian *et al.*, *MRS Bull.* **41**, 775 (2016).
3. L.-C. Zhang and H. Attar, *Adv. Eng. Mater.* **18**, 463 (2016).
4. R. Acharya *et al.*, *Metal. Mater. Trans. B* **45**, 2247 (2014).
5. R. Acharya *et al.*, *Metal. Mater. Trans. B* **45**, 2279 (2014).
6. R. Acharya *et al.*, *Adv. Eng. Mater.* **17**, 942 (2015).
7. R. Acharya and S. Das, *Metal. Mater. Trans. A* **46**, 3864 (2015).
8. M.-T. Nguyen, H.-V. Nguyen and J.-C. Kim, *Vietnam J. Sci. Technol.* **57**, 103 (2019).
9. B. Vrancken *et al.*, *J. Alloys Compd.* **541**, 177 (2012).
10. M. Wang *et al.*, *Prog. Nat. Sci. Mater. Int.* **26**, 671 (2016).
11. G. J. Marshall *et al.*, *JOM* **68**, 778 (2016).
12. B. Song *et al.*, *Front. Mech. Eng.* **10**, 111 (2015).
13. A. Rubenchik *et al.*, *Appl. Opt.* **54**, 7230 (2015).
14. C. Li, Y. B. Guo and J. B. Zhao, *J. Mater. Process. Technol.* **243**, 269 (2017).
15. R. Boyer, E. W. Collings and G. Welsch, *Materials Properties Handbook: Titanium Alloys* (ASM International, Materials Park, OH, 1994), p. 1169.
16. C. Kose and E. Karaca, *Metals* **7**, 221 (2017).
17. R. Dąbrowski, *Arch. Metal. Mater.* **56**, 703 (2011).
18. J. Sieniawski *et al.*, *Titanium Alloys: Advances in Properties Control*, eds. J. Sieniawski and W. Ziaja (IntechOpen, Rijeka, Croatia, 2013), pp. 69–80.
19. W. Xu, M. Brandt and S. Sun, *Acta Mater.* **85**, 74 (2015).
20. E. W. Lui *et al.*, *JOM* **69**, 2679 (2017).

Automatic mantispid egg detection and counting using image nature*

Pei-Ying Yang[†,‡], Chin-Dar Tseng[†], Tai-Lin Huang[§], Chao-Hong Liu[†],
I-Hsing Tsai[†], Wei-Chun Lin[¶], Chin-Shiuh Shieh[†,∥], Shyh-An Yeh[†],
Stephen Wan Leung[†,**] and Tsair-Fwu Lee[†,‡,∥,††]

[†] *Department of Electronics Engineering,*
National Kaohsiung University of Science and Technology,
Kaohsiung 80778, Taiwan, ROC

[‡] *Department of Radiation Oncology,*
Kaohsiung Chang Gung Memorial Hospital
and Chang Gung University College of Medicine,
Kaohsiung 83342, Taiwan, ROC

[§] *Department of Hematology and Oncology,*
Kaohsiung Chang Gung Memorial Hospital
and Chang Gung University College of Medicine,
Kaohsiung 83342, Taiwan, ROC

[¶] *Department of Orthopedics, Kaohsiung Municipal Min-Sheng Hospital,*
Kaohsiung 80276, Taiwan, ROC

[∥] *PhD program in Biomedical Engineering, Kaohsiung Medical University,*
Kaohsiung 80708, Taiwan, ROC
[**] *lwan@ms36.hinet.net*
[††] *tflee@nkust.edu.tw*

Mantispids are small brown bugs about 1.5 cm in length. Mantispid eggs are produced in large quantities, with about 1000 eggs per spawning, and are tiny and densely packed. Traditionally, mantispid eggs are counted manually. However, counting such a large quantity of eggs is difficult. To provide accurate data for researchers, we detail methods to accurately detect and count the number of mantispid eggs using image processing. The following methods were used to count the mantispid eggs: background estimation, morphological image processing, background subtraction, stretching, image thresholding, gray-level transformation, labeling and counting. The results of automated counting were

---

[**,††]Corresponding authors.

*To cite this article, please refer to its earlier version published in the *International Journal of Modern Physics B*, Volume 34, 2040138 (2020), DOI: 10.1142/S0217979220401384.

compared with the results of manual counting. The segmentation results were verified, and the accuracy of the mantispid egg counts was determined to be 100%. This provides a useful resource for mantispid egg counting. The automatic counting system cannot only count mantispid eggs, it can also be used to count other similar insect eggs.

Keywords: Mantispid; automatic counting; background estimation; labeling; counting.

1. Introduction

Mantispids are small brown bugs about 1.5 cm in length [see Fig. 1(a)]. Mantispid larvae are usually parasitic, using the eggs of the *Lycosa* genus or the bodies of young spiders as hosts. Mantispid eggs are produced in large quantities [Fig. 1(b)], at about 1000 eggs per spawning, and are tiny and densely packed. Traditionally, mantispid eggs are counted manually.

However, it is very difficult to count such a large quantity of eggs. To provide accurate data for researchers, we provide detailed methods for detecting and counting mantispid eggs using image processing. These efficient image processing methods enable the accurate counting of the eggs. A flow chart depicting our methods is shown in Fig. 1(c).

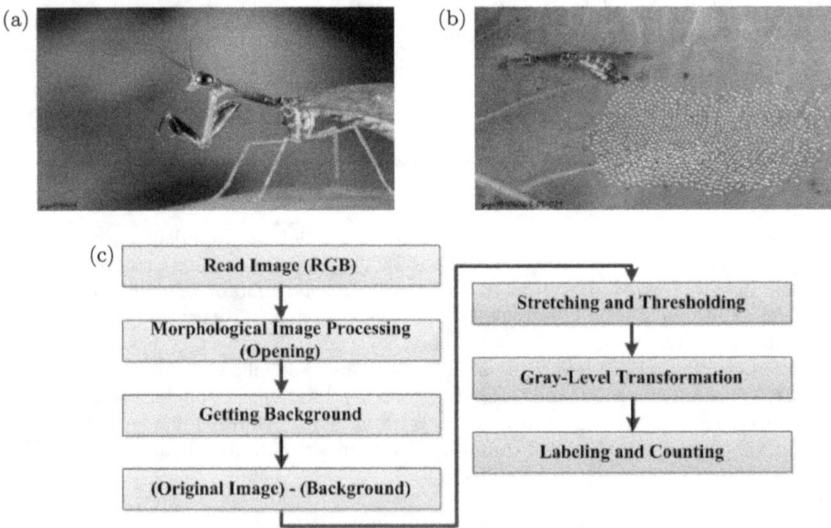

Fig. 1.   Mantispid and flow chart (a) Mantispid adult, (b) Egg laying and (c) Flow chart of methods.

2. Materials and Methods

A clustering method was used with Gaussian mixture approximation to fit the data. For ease of counting, the covariance matrix was assumed to be of the form $\Sigma = \sigma^2\, I$. In each pixel, a cluster is selected as a background process.[1] Measurement of pixel depth is largely accurate under normal circumstances, when the background is simply formed of the furthest pixels, and covers at least T% of the data in the temporal mode. Here, T is the pixel depth of the background,[2] $T = 10$.

Dilation and erosion are two operations that can be performed on images during morphological image processing.[3-5] The dilation of A by B is expressed by the following equation:

$$A \oplus B = \{Z \mid [(\hat{B})_z \cap A] \subseteq A\}, \quad \text{B:a structuring element.} \tag{1}$$

The erosion of A by B is expressed by Eq. (2)

$$A \ominus B = \{Z \mid (B)_z \subseteq A\}. \tag{2}$$

The opening of set A by structuring element B, denoted $A \circ B$, is defined by Eq. (3)

$$A \circ B = (A \ominus B) \oplus B. \tag{3}$$

Background subtraction is also referred to as foreground detection.[6] In the field of image processing and computer vision, a foreground image is extracted for further processing (object recognition, etc.). Background subtraction is a widely used method for detecting moving objects using a static video camera (Fig. 2).

Contrast stretching (usually called normalization) is the process of trying to draw out the desired image by stretching the intensity values within a specified range.[6] We used a method to automatically determine the threshold value for use in image segmentation. The simplest method for threshold

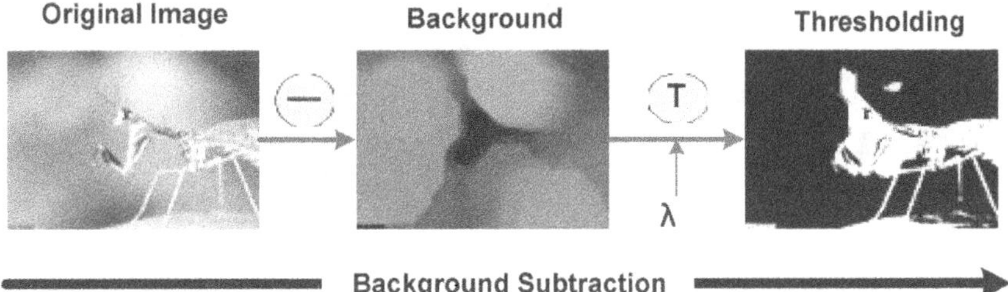

Fig. 2.   Background subtraction (original image-background), T: thresholding.

processing yields either an image with a black pixel, if the image intensity is less than a fixed constant T (i.e., $I_{ij} < T$), or a white pixel if the image intensity is greater than that constant.[7-9] The transformations are described by Eq. (4)

$$s = T(r), \tag{4}$$

where r is the input image and s is the output image. T is a transformation function, which maps each value of r to each value of s.

Labeling and counting are achieved using a sequential algorithm. In the following example, it is assumed that the image is scanned from left to right and from top to bottom. Every pixel is checked for either four- or eight-connectivity, depending on the method chosen. Every pixel is also checked for eight-connectivity for a given region.[10]

3. Result and Discussion

A very small part of the image was segmented from the original image in Fig. 1(b). This is a very small part of the image that contains few eggs [Fig. 3(a)]. The purpose of segmenting the image is to verify the accuracy of counting. Twelve eggs were counted manually, and the same number was counted using image nature. There is a difference of these factors because of the presence of incomplete eggs.

A second small image was segmented from the original image in Fig. 1(b). This image contains more eggs [Fig. 3(b)]. Nineteen eggs were counted using both manual and automated counting, with accuracy being maintained.

Finally, all of the eggs in Figs. 3(c)–3(f) were counted. It was difficult to count all the eggs manually due to the number of eggs, while the automated

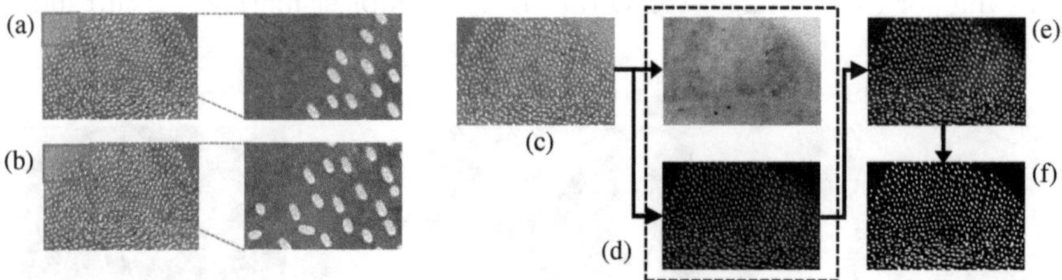

Fig. 3. (Color online) (a) A very small part of the image from Fig. 1(b), (b) a small part of image from Fig. 1(b), (c) original image of Fig. 1(b), (d) getting background and (original image)-(background), (e) stretching and thresholding and (f) gray-level transformation, labeling and counting.

count was 960. According to the count of eggs in the previous segmented image, we believe this result to be valid.

In a previous study, Bouwmans *et al.* reported a widely used approach for background modeling with mixture of Gaussians to detect objects from static cameras.[6] The comparison based on the results shows that our method is beneficial to static image processing. Zhang *et al.* reported that a global multi-level thresholding method for image segmentation is superior to traditional entropy thresholding.[9]

Our approach is to use imaging software to remove the mantispid corpse. Mantispid eggs are white, and the leaves on which they are laid are green. White eggs and green leaves contrast well, which allows image processing with good results. In this study, a cascade method was used for segmentation thresholding. The results of automated counting were compared with the results of manual counting. The results from segmentation counting showed the accuracy of the mantispid egg counting to be 100%. Hence, our system is a good resource for mantispid egg counting.

The automated counting system proposed herein can be used to count the numbers of mantispid eggs and other, similar insect eggs. The performance of the counting task provides a solid foundation for further development of this platform, which could later be extended or used on different systems.

4. Conclusions

The egg counting method proposed herein enabled us to count a large quantity of mantispid eggs with high accuracy to be 100%. We may be able to use the system to count eggs of specific sizes. In addition, our method may be used in a broader context, such as for counting fruits. However, as some insect eggs are arranged in overlapping patterns, our future work will aim develop a method to count overlapping eggs accurately.

Acknowledgements

Pei-Ying Yang and Chin-Dar Tseng contributed equally to this work. The authors would like to thank entomologist Mr. Lin Yi-Hsiung for providing the original photographs. This work was supported by the Minister of Science and Technology, Taiwan, under the Grant Number MOST 107-2221-E-992-014-MY2.

References

1. S. Lu, B. Su and C. L. Tan, *Int. J. Doc. Anal. Recog.* **13**, 303 (2010).
2. M. Vargas *et al.*, *IEEE Trans. Vehicular Technol.* **59**, 3694 (2010).
3. J. Liang, J. Piper and J.-Y. Tang, *Pattern Recog. Lett.* **9**, 201 (1989).
4. A. Downton and D. Crookes, *Electron. Commun. Eng.* **10**, 139 (1998).
5. A. Plaza *et al.*, *Inform. Technol. Govern. Risk Sec.* **40**, 2025 (2002).
6. T. Bouwmans, F. El Baf and B. Vachon, *Recent Patents Comput. Sci.* **1**, 219 (2008).
7. M. Sezgin and B. Sankur, *J. Electron. Image* **13**, 146 (2004).
8. P. K. Saha and J. K. Udupa, *ITPAM* **23**, 689 (2001).
9. Y. Zhang and L. Wu, *Entropy* **13**, 841 (2011).
10. M. B. Dillencourt, H. Samet and M. Tamminen, *J. ACM* **39**, 253 (1992).

Study on the effects of tooth profile design parameters of rotor to performance of vacuum pump*

Van-The Tran[†], Bui Trung Thanh[‡], Banh Tien Long[§] and Hoang Quoc Tuan[¶]

*Department of Mechanical Engineering,
Hung Yen University of Technology and Education,
39 Rd., Hung Yen City, Vietnam*
[†]*vanct4.hut@gmail.com*
[‡]*buitrungthanh@gmail.com*
[§]*long.banhtien@hust.edu.vn*
[¶]*hqtcdt@gmail.com*

Duc Toan Nguyen

Hanoi University of Technology and Science, Hanoi, Vietnam
toan.nguyenduc@hust.edu.vn

The vacuum pump usually used traditional curves such as the circular, cycloidal curves and their combinations to construct tooth profile. However, to increase efficiency and design flexibility for the vacuum pump, a novel rotor tooth profile for Roots rotor of vacuum pumps is proposed. Which is named "CEIEC" tooth profile and orderly composed of five significant segments, a circular arc for tooth tip, an epicycloid curve with variable extension, an involute, an enveloped epicycloid curve and a conjugated circular arc for tooth root. A numerical example is presented to evaluate the performance indices for proposed vacuum pump, including the hermeticity coefficients of the rotor mesh gap and tip gap.

Keywords: CEIEC tooth profile; hermeticity coefficient; roots rotor; vacuum pump.

1. Introduction

The roots pumps are used widely in industrial applications such as the food, medicine and biotechnology. In addition, they can be able to work

---

[†]Corresponding author.
*To cite this article, please refer to its earlier version published in the *International Journal of Modern Physics B*, Volume 34, 2040141 (2020), DOI: 10.1142/S0217979220401414.

with various materials, including low viscosity fluids such as water, very high viscosity fluids such as oil, and even solids. In the operating process of a pump, the tooth profile of Roots rotor (lobe pump) is an important factor for improving performance of the vacuum pump. It permits the Roots rotors to remain meshing with each other.

The tooth profile design for Roots rotor of vacuum pump has received much attention by many researchers. Firstly, Litvin[1] and Tsay[2] proposed a geometric design for tooth profile of a rotor with two lobes using a single circular arc. Meanwhile, a new tooth profile for rotor is developed by combining circular arcs and a conjugated epicycloidal curve.[3] Fang[4] patented an addendum and a dedendum portions for the tooth profiles of rotor by comprising four circular arcs that can improve area efficiency of a vacuum pump. Subsequently, the tooth profile of the rotor is developed with the combination of five circular arcs by Fong *et al.*[5] Previously, Niimura *et al.*[6] patented an addendum tooth profile consisting of a circular arc and an involute to increase pump efficiency. More lately, an extended cycloid curve with a variable trochoid ratio to is applied to improve pump performance by Hwang and Hsieh.[7,8] Kang *et al.*[9,10] developed a new lobe pump rotor profile that used circular and epicycloidal curves that can significantly improve pump performance. Besides, a dynamic mesh method is proposed to provide factors affecting the performance of lobe pumps. A lobe profile design of rotor consisting of a hypocycloid and epicycloid and a manufacturing method are presented by Chiu.[11] Shujun *et al.*[12] presented a tooth profile design of rotor by modifying the traditional involute profile and obtains a new involute profile for improving efficiency of vacuum pump. More recently, Hsieh[13] proposed a new elliptical roulette curve for rotors of the lobe pumps that can achieve the high discharge efficiency and control the level of vibration and noise. Finally, Wu and Tran[14] proposed a mathematical model and method to generate a novel rotor profile for multi-stage vacuum pumps.

This paper proposes a novel tooth profile, named "CEIEC" tooth profile, for Roots type vacuum pumps that is composed of five main curves which comprise orderly one circular arc on the rotor tooth tip, one epicycloid with variable extension quantity, an involute, one enveloped curve by the epicycloid and one conjugated circular arc in the rotor tooth root. Numeral example is presented to illustrate the effect of tooth profile design parameters to increase the performance of vacuum pump.

2. Mathematical Models for Generating Tooth Profiles of the Roots Rotors

As shown in Fig. 1, the CEIEC tooth profile of Roots rotor is comprised by five significant curves, wherein the tooth profile is in axial symmetry and only one-half of the tooth profile requires definitions. The tip segment of the Roots rotor $A_1 B_1$ is a circular arc. The circular arc is used for ensuring the hermeticity against the chamber with its center coinciding with the rotation center of rotor. The segment $B_1 C_1$ is an epicycloid with variable extension. The epicycloid is used for improving adjustable flexibility of the rotor tooth profile design. The segment $C_1 D_1$ is an involute curve. The involute curve is used for preventing possible rotor surface wear on the gas-sealing band across the rotor pitch circle. The segment $D_1 E_1$ is a corresponding hypocycloid enveloped by the epicycloid $B_1 C_1$. And the segment $E_1 F_1$ is the tooth root circular arc meshed with the segment $A_1 B_1$. The position vector of the tooth profile of Roots rotor 1 are defined as below:

The position vector of CEIEC tooth profile of Roots rotor comprised by five significant curves is represented in the coordinate system S_1:

(1) The circular arc $A_1 B_1$:

$$
\mathbf{r}_{I_1}^{(A_1 B_1)}(\tau) =
\begin{cases}
x_{I_1}^{(A_1 B_1)}(\tau) = r_a \cos \tau \\[2mm]
y_{I_1}^{(A_1 B_1)}(\tau) = r_a \sin \tau
\end{cases}
, \quad (0 \leq \tau \leq \theta_1). \tag{1}
$$

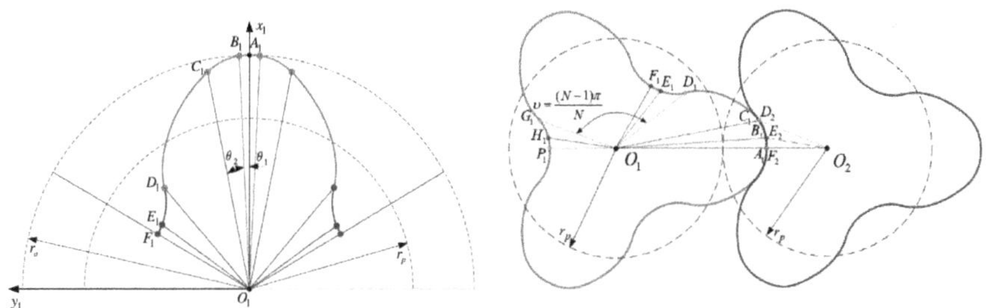

Fig. 1. (Color online) The CEIEC tooth profile of the proposed Roots rotors.

(2) The extended epicycloid curve B_1C_1:

$$\mathbf{r}_{I_1}^{(B_1C_1)}(\tau) = \begin{cases} x_{I_1}^{(B_1C_1)}(\tau) = (r_p + r_e)\cos(\tau + \theta_1) \\ \qquad + d\cos\left(\tau + \dfrac{r_p}{r_e}\tau + \theta_1\right) \\ y_{I_1}^{(B_1C_1)}(\tau) = (r_p + r_e)\sin(\tau + \theta_1) \\ \qquad + d\sin\left(\tau + \dfrac{r_p}{r_e}\tau + \theta_1\right) \end{cases}, \quad (0 \le \tau \le \theta_2).$$

$$\tag{2}$$

(3) The involute curve C_1D_1:

$$\mathbf{r}_{I_1}^{(C_1D_1)}(\tau)$$

$$= \begin{cases} x_{I_1}^{(C_1D_1)}(\tau) = (r_p + \tau\cos\alpha_p)\cos\left(\dfrac{\pi}{2N} - \dfrac{\tau}{r_p\sin\alpha_p}\right) \\ \qquad - \tau\cos\alpha_p\cot\alpha_p\sin\left(\dfrac{\pi}{2N} - \dfrac{\tau}{r_p\sin\alpha_p}\right) \\ y_{I_1}^{(C_1D_1)}(\tau) = \tau\cos\left(\dfrac{\pi}{2N} - \dfrac{\tau}{r_p\sin\alpha_p} - \alpha_p\right)\cot\alpha_p \\ \qquad + r_p\sin\left(\dfrac{\pi}{2N} - \dfrac{\tau}{r_p\sin\alpha_p}\right) \end{cases}, \quad (-u \le \tau \le u).$$

$$\tag{3}$$

(4) The hypocycloid enveloped by the epicycloid D_1E_1. The extended epicycloid curve B_1C_1 is a conjugate extended hypocycloidal curve D_2E_2 that is identical dedendum curves G_1H_1, as shown in Fig. 1. Therefore, the curve D_1E_1 can be derived by rotating G_1H_1 with an angle of $\vartheta = (N-1)\pi/N$ counterclockwise about z-axis:

$$\mathbf{r}_{I_1}^{(D_1E_1)}(\tau) = \begin{cases} x_{I_1}^{(D_1E_1)}(\tau) = (r_p - r_e)\cos\left(\dfrac{\pi}{N} - \tau - \theta_1\right) \\ \qquad - d\cos\left(\dfrac{\pi}{N} - \tau + \dfrac{r_p}{r_e}\tau - \theta_1\right) \\ y_{I_1}^{(D_1E_1)}(\tau) = (r_p - r_e)\sin\left(\dfrac{\pi}{N} - \tau - \theta_1\right) \\ \qquad - d\sin\left(\dfrac{\pi}{N} - \tau + \dfrac{r_p}{r_e}\tau - \theta_1\right) \end{cases}. \tag{4}$$

(5) The circular arc E_1F_1. The addendum circular arc A_1B_1 is conjugated dedendum circular arc E_2F_2 that is identical dedendum curves H_1P_1, as shown in Fig. 1. Therefore, the dedendum circular arc E_1F_1 can be obtained by rotating H_1P_1 with an angle of $\vartheta = (N-1)\pi/N$ counterclockwise about z-axis:

$$
\mathbf{r}_{I_1}^{(E_1F_1)}(\tau) = \begin{cases} x_{I_1}^{(E_1F_1)}(\tau) = 2r_p\cos(\vartheta-\tau) + r_a\cos(\vartheta+3\tau) \\ \\ y_{I_1}^{(E_1F_1)}(\tau) = 2r_p\sin(\vartheta-\tau) + r_a\sin(\vartheta+3\tau) \end{cases}. \tag{5}
$$

3. Definitions of Performance Indices for Roots Rotor Tooth Profiles of Vacuum Pump

To evaluate the gas leakage and hermeticity for Roots rotor's tooth profiles, two performance indices are proposed and defined. One is the hermeticity coefficient of rotor mesh gap which is used to evaluate the gas seal characteristic between two engaged rotors, and another is the hermeticity coefficient of rotor tip gap which is used to access the hermeticity between the tooth tip and chamber in a single tooth mesh process. The calculation model of seal line length based on discrete points of the rotor's tooth profiles has considered the gaps in a certain time, as shown in Fig. 2. The hermeticity coefficients between two rotors and the rotor tip against the chamber are defined as:

$$
\kappa_{st} = \frac{\frac{1}{k}\sum_{i=1}^{k} L_i^{-1}}{\sum_{i=1}^{k}\left|\mathbf{r}_{i+1}^{(1)} - \mathbf{r}_i^{(1)}\right|}, \quad \kappa_{sc} = \frac{\left|\mathbf{r}_{i+1}^{(1)} - \mathbf{r}_i^{(1)}\right|}{r_c\frac{\pi}{N_1} + \sum_{i=1}^{k}\left|\mathbf{r}_{i+1}^{(1)} - \mathbf{r}_i^{(1)}\right|}, \tag{6}
$$

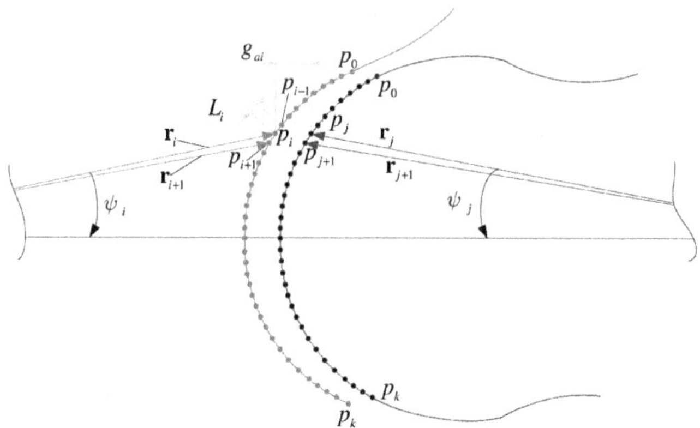

Fig. 2. (Color online) Model for calculating the length of seal line.

where L_i is the seal line length between two points p_{i-1} and p_i, g_{ai} is the normal gap at the same point p_i on the driving and driven rotors (Fig. 2), $0 \leq g_{ai} \leq g_{a\,\max}$ and $g_{a\,\max}$ is the maximum normal gap, it is chosen to be equal to 0.2 mm, and r_c is the radius of the chamber.

4. Numeral Example

The influences of design parameters, top tooth arc angle (θ_1), exponent of the variable extension function (n) and involute parameter (u) on the hermeticity coefficients of the rotor mesh gap and the tip gap for proposed rotor (CEIEC tooth profile) are investigated. The basic data of the Roots rotor and tooth design parameters are given in Table 1.

The effects of each parameter on the hermeticity coefficients are considered in ideal condition. The varying values of design parameters are considered in three cases: Case 1 (θ_1), Case 2 (n) and Case 3 (u), as listed in Table 2. The hermeticity coefficients with corresponding variation of the design parameters are presented in Table 2 and Figs. 3–5. The hermeticity coefficient of the rotor mesh gap is much sensitive to the variation of involute parameter, u, but it is not much sensitive to the variation of top tooth arc angle, θ_1, and exponent of the variable extension function, n. In detail, the hermeticity coefficient, $\kappa_{st} = 0.110$, with involute parameter, $u = 9\ mm$, has increased to $\kappa_{st} = 0.136$ and $\kappa_{st} = 0.123$ when the involute parameter is reduced to $u = 6\ mm$ and increased to $u = 12\ mm$, respectively. In contrast, the hermeticity coefficient of the rotor mesh gap between two rotors, $\kappa_{st} = 0.110$, with top tooth arc angle, $\theta_1 = 4°$, has not much changed when the top tooth arc angle is reduced to $\theta_1 = 2°$ and increased to $\theta_1 = 6°$, respectively. The hermeticity coefficient of the rotor mesh gap is a slightly sensitive to the variation of the exponent of the variable extension function, n. However, the hermeticity coefficient of the tip gap is much sensitive to the variation of design parameters.

Table 1. Basic data of the root rotor.

Roots rotor		Roots rotor	
Number of rotor lobes (N)	3	Normal pressure angle (α_p)	36.43°
Normal module (m)	22	Center distance (E_c)	66 mm
Pitch radius (r_p)	33.0 mm	Involute parameter (u)	9.00 mm
Outer radius (r_a)	44.88 mm	Top tooth arc angle (θ_1)	4.0°
Normal gap (g_a)	0 μm	Exponent of the variable extension function (n)	2

Table 2. Utilization coefficient of area and hermeticity coefficients with the variation of design parameters.

Design parameters	Case 1			Case 2			Case 3		
	(a)	(b)	(c)	(a)	(b)	(c)	(a)	(b)	(c)
Top tooth arc angle (θ_1)	2°	4°	6°	4°	4°	4°	4°	4°	4°
Exponent of the variable extension function (n)	2	2	2	2	3	4	2	2	2
Involute parameter (u)	9	9	9	9	9	9	6	9	12
Utilization coefficient of area (η)	0.440	0.439	0.438	0.439	0.438	0.437	0.436	0.439	0.443
Hermeticity coefficient of the rotor mesh gap (κ_{st})	0.110	0.110	0.118	0.110	0.119	0.116	0.136	0.110	0.123
Hermeticity coefficient of the tip gap (κ_{sc})	0.139	0.160	0.181	0.160	0.177	0.182	0.169	0.160	0.148

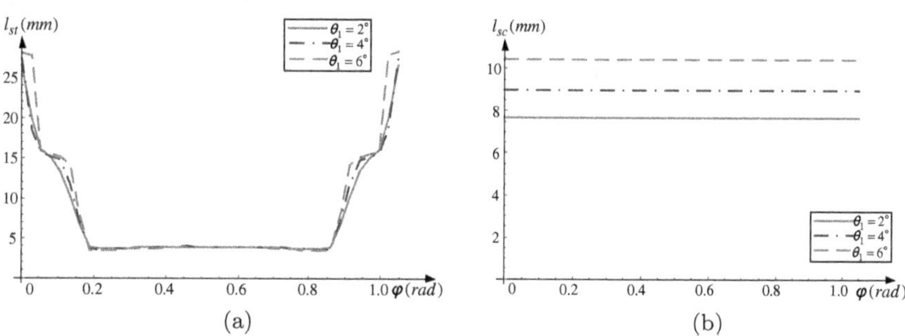

Fig. 3. (Color online) The length of seal line between (a) two rotors and (b) rotor and chamber with the variation of top tooth arc angle (θ_1).

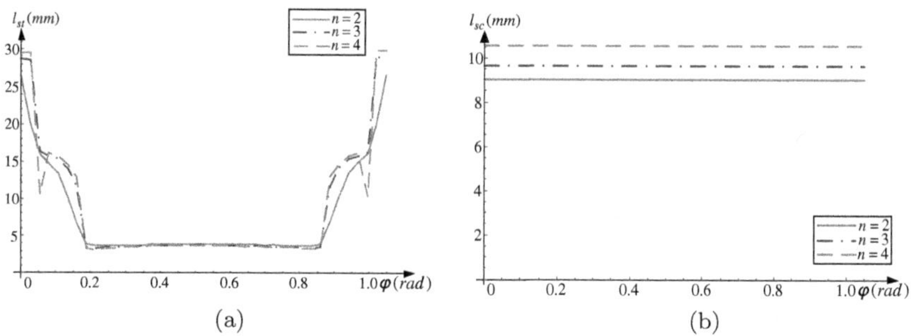

Fig. 4. (Color online) The length of seal line between (a) two rotors and (b) rotor and chamber with the variation of exponent of variable extension function (n).

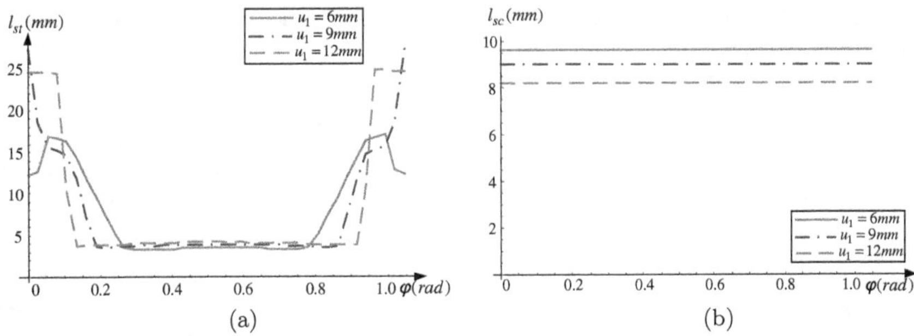

Fig. 5.   (Color online) The length of seal line between (a) two rotors and (b) rotor and chamber with the variation of involute coefficient (u).

5. Conclusions

In this paper, a mathematical model for CEIEC of tooth profiles of Roots rotor of the vacuum pump have constructed. A computer program is built to simulate and compare the hermeticity coefficients for the CEIEC tooth profile of Roots rotors. Besides, the sensitivities of design parameters to the hermeticity coefficients are investigated. It is found that the hermeticity coefficient of the rotor mesh gap is not much sensitive to the change of top tooth arc angle, exponent of the variable extension function. But it is much sensitive to the change of the involute parameter. And the hermeticity coefficient of the tip gap is also much sensitive to the change of the design parameters.

Acknowledgments

The authors would like to thank for the subsidy of the Outlay B2019-SKH-03 given by the Ministry of Education and Training, Vietnam, to assist us in successfully finishing this special research.

References

1. F. L. Litvin and A. Fuentes, *Gear Geometry and Applied Theory*, 2nd edn. (Cambridge University Press, United Kingdom, 2004).
2. C. B. Tsay, *J. Mach.* **13**, 129 (1987).
3. D. Vecchiato, A. Demenego, J. Argyris and F. L. Litvin, *Computer Methods Appl. Mech. Eng.* **190**, 2309 (2001).
4. H. S. Fang, Rotor profile for a Roots vacuum pump, U.S. Patent 5,152,684 (1992).
5. P. Y. Wang, Z. H. Fong and H. S. Fang, *J. Mech. Eng. Sci. C* **216**, 225 (2002).

6. Y. Niimura, R. Kikuta and K. Usui, Two-Shaft type rotary machine having a tip circle diameter to shaft diameter within a certain range, US Patent No. 4943214 A (1990).
7. Y. W. Hwang and C. F. Hsieh, *Proc. Inst. Mech. Eng. C* **220**, 1375 (2006).
8. C. F. Hsieh and Y. W. Hwang, *Math. Computer Modelling* **48**, 19 (2008).
9. Y. H. Kang, H. H. Vu and C. H. Hsu, *J. Mech.* **28**, 229 (2012).
10. Y. H. Kang and H. H. Vu, *J. Mech. Sci. Technol.* **28**, 915 (2014).
11. H. C. Chiu, *J. Technol.* **9**, 13 (1994).
12. S. Wang, H. Li, Y. Zhao, H. Liu and L. Wei, The improvement study of involutes profile type rotor profile in Roots vacuum pump, *2011 IEEE International Conference on New Technology of Agricultural Engineering* (Zibo, China, 2011), pp. 251–253.
13. C. F. Hsieh, *Mech. Mach. Theory* **87**, 70 (2015).
14. Y. R. Wu and V. T. Tran, *Mech. Mach. Theory* **128**, 457 (2018).

Comparative study of low-frequency vibrations assigned to a workpiece in EDM and PMEDM*

Quang Dung Le

Faculty of Mechanical Engineering,
Hung Yen University of Technology and Education, Vietnam

Huu Phan Nguyen[†]

Faculty of Mechanical Engineering,
Hanoi University of Industry, Vietnam
phanktcn@gmail.com

Tien Long Banh and Duc Toan Nguyen

School of Mechanical Engineering,
Hanoi University of Science and Technology, Vietnam

This study deals about the influence of vibrations incorporated into a workpiece during powder-mixed electrical discharge machining (PMEDM) on quality measures such as material removal rate (MRR), surface roughness (R_a) and microhardness. It has been found that the low-frequency vibration incorporated into the workpiece positively affects the processing efficiency of electrical discharge machining (EDM) and PMEDM. However, the effect of low-frequency vibration in PMEDM has been better than EDM. The higher vibration frequency significantly improves the MRR and R_a in PMEDM. The MRR has been improved by 95.89% and with lower R_a of 63.2% in PMEDM. The hardness of the machined surface after PMEDM using titanium powder mixed in dielectric liquid was increased approximately two times as compared with conventional EDM.

Keywords: EDM; PMEDM; vibration; MRR; R_a.

[†]Corresponding author.

*To cite this article, please refer to its earlier version published in the *International Journal of Modern Physics B*, Volume 34, 2040145 (2020), DOI: 10.1142/S0217979220401451.

1. Introduction

The optimal selection of appropriate material types and sizes should be mixed into the dielectric liquid in powder-mixed electrical discharge machining (PMEDM). It contributes to improving the productivity and quality of the surface layer compared to standard electrical discharge machining (EDM).[1-3] The literatures on the effects of vibrations in EDM have shown that vibrations can have a very positive influence in PMEDM process. It can improve various limitations in PMEDM such as the uniformity of the powder particles in the dielectric liquid and the density of the powder particles in the discharge gap. The new powder selection with dielectric medium in the discharge gap can also be performed.[4,5] Only a limited number of research works exist about PMEDM with vibration. These research results only study a particular type of powder material, electrode material and workpiece material.[6] The new research works deal with different types of powders, electrodes and workpieces in the PMEDM process. It is very essential and contributes to the promotion of PMEDM process with vibration in practical scenarios. In this study, the effects of low-frequency vibrations on specific quality characteristics have been evaluated in PMEDM process. Material removal rate (MRR), SR and HV are used as performance measures in this study. The considerable low-frequency vibration has been incorporated into the workpiece for the experimental investigation.

2. Experimental Methodology

A CNC CM323C die-sinking machine (CHMER, Taiwan) was used for the performing machining investigation with copper as tool electrode with dimensions of 12 mm. A vibration unit (Model: Exciter 4824, Brüel and Kjær, Denmark) was connected with the fixture of workpiece with EDM machine under the range 0–2 KHz. The amplitude (a) of the vibrations for the chosen frequency was 0.75 mm. The experimental diagram is shown

Table 1.   Input of the process parameters.

Adjustable parameters	Symbol	Unit	Value
Voltage	U	V	60
Pulse on time	T_{on}	μs	25
Pulse off time	T_{of}	μs	12.5
Current	I	A	8A
Powder concentration	C_0	g/liter	4
Frequency	F	Hz	0; 200; 400; 600

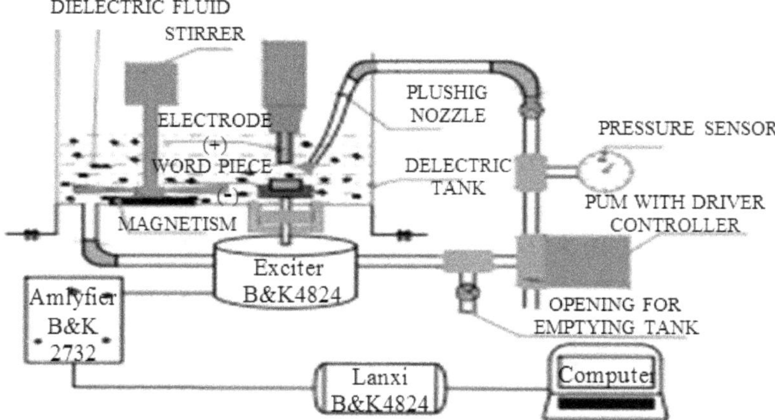

Fig. 1. Schematic experimental chart depicting the vibration set up and data logger.

in Fig. 1. SKD61 steel was chosen as workpiece size of $20 \times 20 \times 10$ mm. The titanium powder was 45 μm size and mixed with dielectric medium.

3. Results and Discussion

3.1. *Effect of frequency on material removal rate (MRR)*

Higher MRR in PMEDM has been found by incorporating vibrations with the workpiece in EDM and PMEDM, which significantly improved MRR, as shown in Fig. 2. The MRR in PMEDM is three times larger as compared with the MRR in EDM without vibration. It shows the positive effect of titanium powder mixed into the dielectric liquid. An increase in F led to an increase in MRR in EDM and PMEDM. MRR in EDM with vibration had a maximum increase of 34.94% with $F = 600$ Hz compared to EDM alone. MRR in PMEDM under $F = 400$ Hz had a maximum increase of 95.89% compared to PMEDM without vibration. It shows that a higher F resulted in better MRR in PMEDM with vibration. The incorporating vibrations into the workpiece led to MRR in PMEDM being increased very sharply as compared with EDM. MRR in PMEDM with vibration was higher than MRR in EDM with vibration approximately 4.6 times higher at $F = 200$ Hz.

3.2. *Effect of frequency on surface roughness (R_a)*

The change in R_a in EDM and PMEDM with the change in F is shown in Fig. 3. In EDM and PMEDM without vibration, R_a in PMEDM is

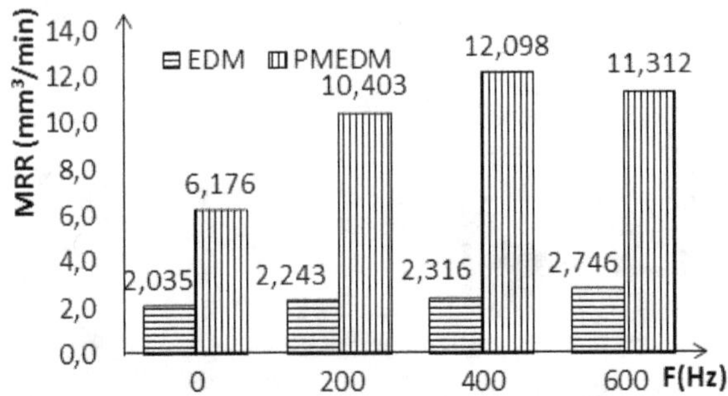

Fig. 2.   Effect of low-frequency vibration to the material removal in EDM and PMEDM.

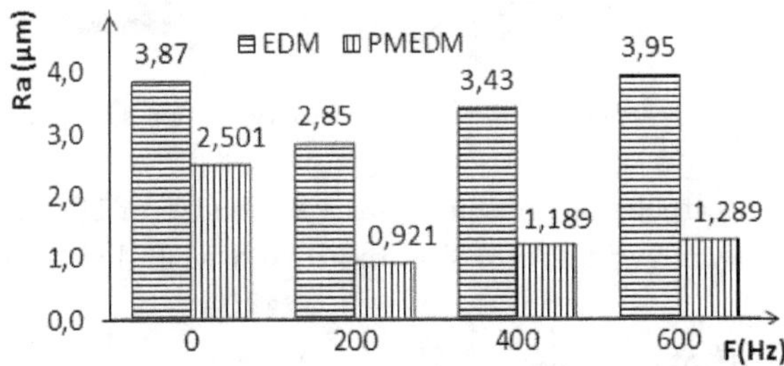

Fig. 3.   Effect of low-frequency vibration to the surface roughness in EDM and PMEDM.

approximately 35.4% smaller than conventional EDM. The lower R_a in PMEDM was observed with vibration-assisted powder-mixed EDM. The R_a in EDM under $F = 200$ Hz was reduced by a maximum of 63.2% and 26.4%, respectively as compared with $F = 400$ Hz and 600 Hz. However, the increase in R_a in PMEDM is lower than the vibration-assisted EDM. At $F = 600$ Hz, the R_a in vibration-assisted EDM was increased by 2.1%. Nevertheless the R_a in PMEDM with vibration was reduced by 48.5% than the conventional EDM. This shows that vibration has a very positive effect on R_a in PMEDM.

3.3. *Effect of frequency on microhardness of surface (HV)*

Figure 4 shows that the hardness of the surface layer after EDM and PMEDM was not significantly affected by vibration. The hardness of the surface layer after machining has been strongly influenced by the mixing powder in the dielectric liquid.[7,8] The hardness of the surface after PMEDM was improved as compared with EDM. The HV in PMEDM was increased

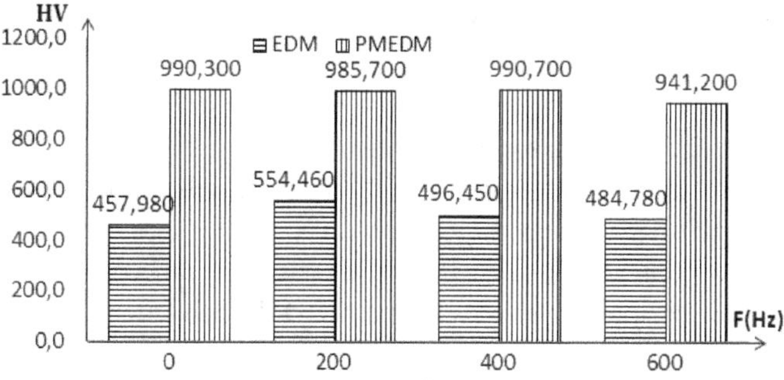

Fig. 4. Effect of low-frequency vibration to the hardness of surface in EDM and PMEDM.

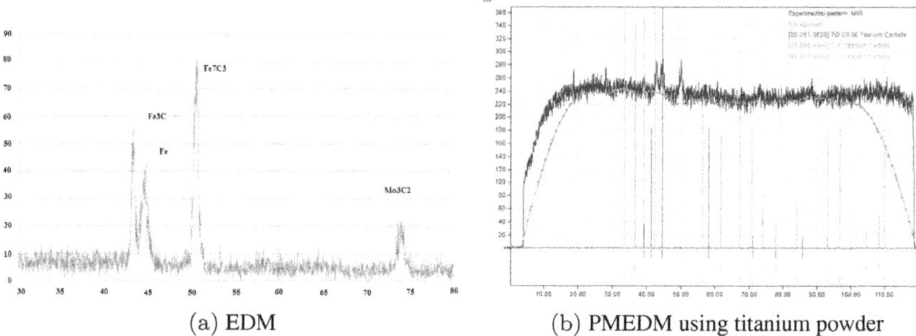

(a) EDM   (b) PMEDM using titanium powder

Fig. 5. (Color online) Phase organization of machined surface after EDM and PMEDM.

by approximately two times as compared with EDM due to the melting and evaporation of titanium powder mixed in the dielectric liquid caused by electric sparks. This has entered the machining surface layer. The phase organization of the machined surface after EDM and PMEDM is shown in Fig. 5. The surface layer after PMEDM has the appearance of a TiC phase organization owing to the carbon in the steel and the carbon generated by the spark splitting the oil dielectric liquid which has been combined with titanium, melted and evaporated during the PMEDM process.

4. Conclusion

In this study, an attempt has been made to analyze the effects of frequency in EDM and PMEDM process on machining SKD61 steel. From the experimental investigations, the following conclusions have been made:

- The titanium powder mixed in EDM can significantly enhance the MRR, R_a, and HV in PMEDM process.
- Low-frequency vibration assisted the EDM and PMEDM, which can improve the quality measures.
- The surface topography is better enhanced by vibrations, which is due to the ability of producing tiny craters.
- Further research studies in PMEDM with vibration should aim to improve the working mechanism of PMEDM with vibration by simulation methods.

Further research studies in PMEDM with vibration will include PMEDM using other powders and will clarify the working mechanism of PMEDM with vibration by simulation methods.

Acknowledgments

This research is funded by the Vietnam National Foundation for Science and Technology Development (NAFOSTED) under Grant Number "107.01–2017.303".

References

1. A. Y. Joshi and A. Y. Joshi, *Heliyon* **5**, e02963 (2019).
2. C. A. O. Luzia *et al.*, *Int. J. Adv. Manuf. Technol.* **103**, 15 (2019).
3. S. Jeavudeen, H. S. Jailani and M. Murugan, *Int. J. Mach. Machinability Mater.* **22**, 47 (2020).
4. K. P. Maity and M. Choubey, *Surf. Rev. Lett.* **26**, 1 (2019).
5. T. Muthuramalingam, *J. Clean. Prod.* **238**, 117894 (2019).
6. J. Huo *et al.*, *Mater. Res. Express* **6**, 086507 (2019).
7. T. Muthuramalingam, *Measurement* **131**, 694 (2019).
8. T. Muthuramalingam *et al.*, *Mater. Res. Express* **6**, 126503 (2018).

Repetitive recycling of 3D printing PLA filament as renewable resources on mechanical and thermal loads[*]

Jeong-Hyo Hong[†] and Tianyu Yu

*Major of Materials Engineering, Korea Maritime and Ocean University,
Busan 49112, Republic of Korea*
[†]*jhhong@kmou.ac.kr*

Soo-Jeong Park

*Department of Mechanical and Aerospace Engineering,
Naval Postgraduate School, Monterey, CA 93943, USA*

Yun-Hae Kim[‡]

*Department of Ocean Advanced Materials Convergence Engineering,
Korea Maritime and Ocean University, Busan 49112, Republic of Korea*
yunheak@kmou.ac.kr

3D printing technology has emerged as a high value-added industry with high
efficiency that has dramatically broken away from the existing material and
manufacturing industry's human-based production system. These technologies,
ranging from small parts to large structures, are rapidly developing due to
the challenge of various filament materials, whereas there are significant con-
cerns about waste filler materials, and complimentary research is needed to
improve them. Polylactic acid (PLA), the representative polymer of 3D printing, contributes to minimizing environmental risks. However, although thermoplastic PLA has excellent reversible properties for heat in terms of sustainable
resources, it is degraded as a low value-added material afterward. Therefore,
in this study, the effect of repetitive recycling on the mechanical and thermal
properties of PLA filaments was analyzed to verify and re-evaluate PLA as a
renewable resource. As a result, recycled PLA has decreased tensile and flexural strength by up to 69% and 53%, respectively, compared to initial neat PLA
with the increase of the number of repetitive recycling, and this demonstrates
the change in the thermal properties of recycled PLA.

[‡]Corresponding author.
[*]To cite this article, please refer to its earlier version published in the *International Journal of Modern Physics B*, Volume 34, 2040147 (2020), DOI: 10.1142/S0217979220401475.

Keywords: Polylactic acid (PLA); 3D printing; polymer recycling; mechanical property; thermal property; renewable material.

1. Introduction

Fused deposition modeling (FDM), a typical 3D printing type, uses molten thermoplastic filaments to manufacture parts and structures. In this process, a large amount of plastic waste-pickers such as filler material supporting shape, residual waste through print nozzle and remaining pieces is generated. These waste materials ultimately reveal the technical limitations of 3D printing, and act as a major factor that lowers efficiency of rational consumption of raw materials compared to productivity.[1] Recently, various advanced studies on recycled materials and their processes have been reported as a part of post improvement. Thermoplastic Polylactic acid (PLA), which is widely used as a 3D printing material, has the characteristics of thermally reversible shape change, and its utilization value is highly appreciated as a part of environmentally-friendly biomaterials.[2] According to the results of advanced studies, neat PLA showed a relatively lower mechanical strength of up to 80% than recycled PLA under tensile loads. This means that PLA filament is vulnerable to damage caused by general environmental factors such as moisture and heat, and initial irreversible damage that occurs before recycling has a limit to strength recovery in the thermal recycling process.[1] Accordingly, various pre- and post-treatment processes have been proposed to secure reusability and sustainability of recycled PLA, such as coating Polydopamine (PDA) on the surface of recycled PLA or blending recycled PLA and neat PLA.[3,4] These studies, however, have resulted in one or two recycling cycles, and the methodological approach proposed based on these results is only limited to the one-time strengthening of mechanical strength. Therefore, in this study, repetitive recycling was carried out to secure the effect of thermal PLA recycling process on mechanical strength characteristics and reasonable recycling limiting value. The failure behavior of PLA was also derived through mechanical and thermal properties as the recycling cycle increased.

2. Experimental Works

2.1. *Recycled PLA polymer and recycling process*

3D printing PLA filament of 1.75 mm diameter was supplied by 3DP200PBW-QR (Shindoh, South Korea), with a density of 1.06 g/cm^3, melting temperature (T_m) of 160°C and glass transition temperature (T_g)

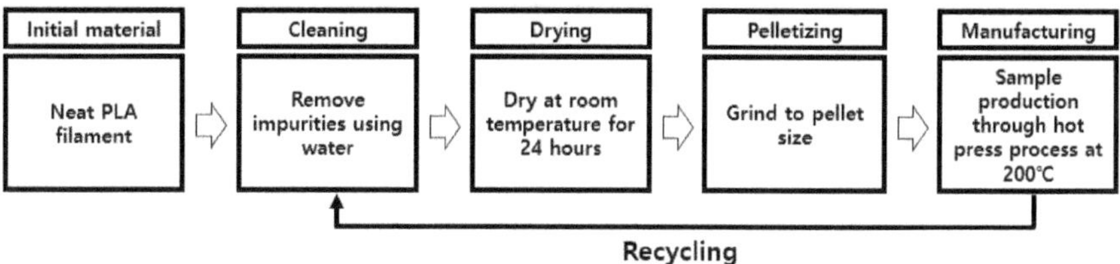

Fig. 1. Repetitive recycling process on PLA filament.

of 80°C. Repetitive recycling defined washing, drying, pelletizing, and manufacturing steps as one cycle, as shown in Fig. 1, and these steps were repeated up to 3 times. Each cycle is as follows; R1: PLA recycled once, R2: PLA recycled twice, R3: PLA recycled 3 times.

2.2. *Evaluation of mechanical and thermal properties*

DSC thermal analysis is intended to demonstrate the change in polymorphic behavior due to repeated thermal recycling processes and its effect on the failure of PLA. The DSC, DSCN-650 (SCINCO, Republic of Korea) was performed in a temperature range of 200°C from 40°C, heating and cooling rates of 10°C/min using Ar gas. To analyze the performance of PLA through repetitive recycling, mechanical analyses was performed under tensile and flexural loads using a KDMT-136 UNIVERSAL TESTING MACHINE (Kyung Do, Republic of Korea), and these were based on ASTM D 638 and ASTM D 790, respectively.

3. Results and Discussion

3.1. *DSC thermal analysis*

The DSC results of the recycled PLA are illustrated in Fig. 2. The T_g of the as-received PLA is approximately 56.36°C, while the T_g of R1, R2 and R3 recycled PLA demonstrate different degrees of decline. The higher T_g, generally, reflects the higher crystalline proportion; the higher crystallinity, on the other hand, requires higher temperature to loosen the restricted amorphous chain segments. The crystallinity of R1, R2 and R3 recycled PLA is checked, to be reduced by T_g comparison. The cold crystallization of the as-received PLA exhibits quite a broad exothermic peak with relative a low crest value. The cold crystallization peak became narrowed and shifted to the low-temperature side with the times of recycle. The lowered cold

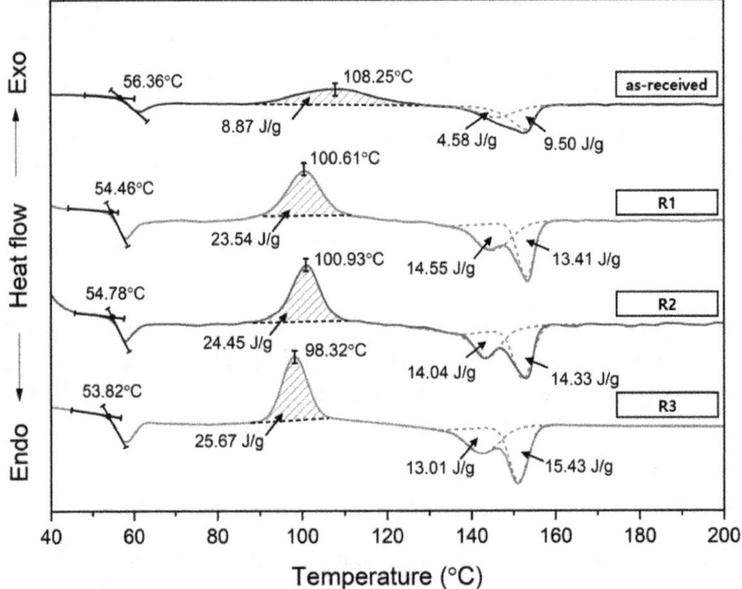

Fig. 2.   (Color online) DSC curves of recycled PLA.

crystallization temperature contributes to the increased proportion of the disordered α phase, which may contribute to the decline to mechanical properties. Also, the more distinguished cold crystallization behavior in recycled PLA indicates the lower content of structure crystallinity, and may induce the mechanical property deterioration and dimensional fluctuation.

3.2. *Fracture behavior under mechanical loadings*

As shown in Fig. 3, a tendency to decrease in mechanical strength was commonly observed with the increase in the number of repetitive recycling under tensile and flexural loads. In Fig. 3(a), recycled PLA maintained tensile strength corresponding to about 63.1% and 60.9%, respectively, of as-received PLA in R1 and R2, and had a loss of tensile strength of about 69% in R3. In Fig. 3(b), flexural strength degreased in the range of about 20% for each cycle in repetitive recycling. As mentioned above, the changes in the thermal properties of recycled PLA ultimately are involved in mechanical performance, and support the possibility of formation of structurally nonuniform regions during the thermal reversible reactions adopted as the principle of PLA recycling. In other words, the nonuniform area

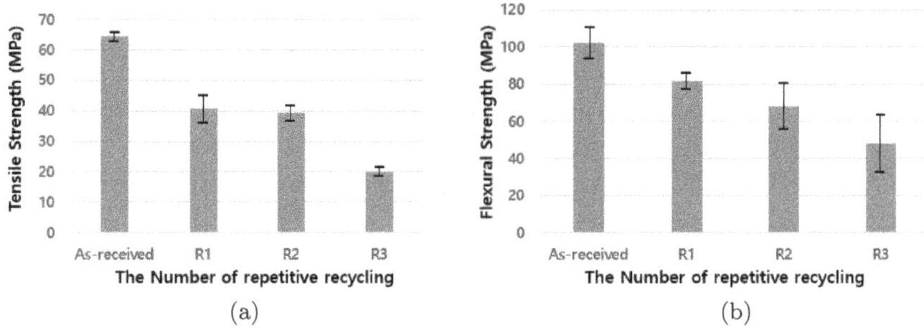

Fig. 3. Comparison of (a) tensile and (b) flexural strength according to the number of repetitive recycling.

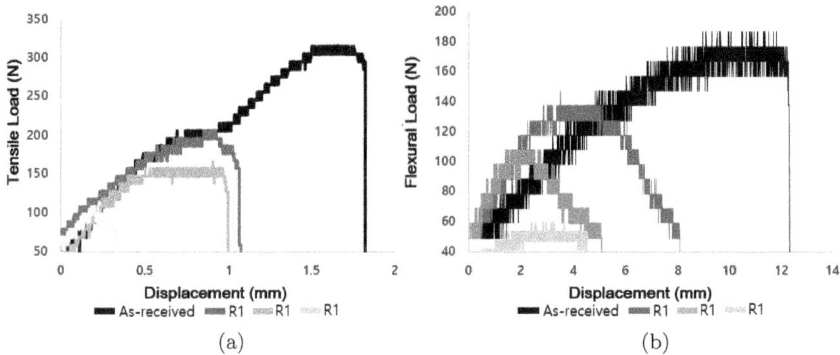

Fig. 4. Load-displacement curves according to the number of repetitive recycling under (a) tensile and (b) flexural loadings.

accumulated in the thermally modified PLA acted as a micro void formed at the interface of a relatively weak nonuniform area due to temperature difference during curing reaction or defects such as impurities in the cured PLA. As a result, it is considered that these factors promoted the failure of repetitively recycled PLA, and the mechanical strength gradually decreased.

Figure 4 shows the load-displacement curves for tensile and flexural test results. The fracture toughness decreased with increasing number of repetitive recycling. Also, the stiffness and tensile modulus tended to be similar under all conditions. The fracture pattern of flexural samples is distinctively changed that the as-received PLA showed the sudden breakage while the recycled PLA demonstrate the progressive failure.

4. Conclusions

(1) Repetitive recycled PLA has a relatively lowered T_g and cold crystallization temperature compared with neat PLA, which resulted in a decrease in fracture toughness against external loads due to a decrease in crystallinity. Particularly, these thermal properties showed the same tendency as in tensile load.

(2) As the number of repetitive recycling increased, the mechanical strength under tensile and flexural loads gradually decreased, and the decreased strength rate was commonly high in the first recycling. This is because the thermally modified PLA acted as a cause of formation of a risk factor for fracture.

(3) It was difficult to specify the reasonable recycling limiting value of PLA with up to three cycles, it is considered that it is necessary to maintain the inherent thermal properties of PLA to improve the decreased strength rate.

References

1. J. H. Hong *et al.*, *Mod. Phys. Lett. B* **33**, 1940025 (2019).
2. M. A. Cuiffo *et al.*, *Appl. Sci.* **7**, 579 (2017).
3. D. Tanney, N. A. Meisel and J. Moore, Investigating material degradation through the recycling of PLA in additively manufactured parts, in *Proc. 28th Annual Int. Solid Freeform Fabrication Symposium — An Additive Manufacturing Conference* (2017), pp. 519–531.
4. X. Tian *et al.*, *J. Clean. Prod.* **142**, 1609 (2017).

Investigations of the characteristics and chemical bonds of Al and F co-doped ZnO films as a function of F_2 flow rate[*]

Guo-Xiang Peng

School of Information and Engineering,
Jimei University, Fujian 361021, P. R. China
gxpeng@jmu.edu.cn

Fang-Hsing Wang[†] and Chun-Chi Lin[‡]

Graduate Institute of Optoelectronic Engineering,
National Chung Hsing University, Taichung 402, Taiwan
[†]*fansen@dragon.nchu.edu.tw*
[‡]*chi90296@hotmail.com*

Cheng-Fu Yang[§]

Department of Chemical and Materials Engineering,
National University of Kaohsiung, Kaohsiung 811, Taiwan
Department of Aeronautical Engineering,
Chaoyang University of Technology, Taichung 413, Taiwan
cfyang@nuk.edu.tw

In this study, we investigate the effect of fluorine (F_2) flow rate on the crystal characteristics and chemical bonds of Al-Doped ZnO (AFZO) Films. At first, 1 wt% Al_2O_3 was mixed with 99 wt% ZnO to form the Al-doped ZnO ceramic target, the radio frequency (RF) magnetron sputtering method was used to deposit the AFZO films by introducing different flow rates of F_2 gas using $F_2/(Ar + F_2)$ mixing atmosphere during the deposition process. F_2 flow rates ($F_2/(Ar + F_2)$) were changed in the range of $0 \sim 0.25\%$, and F_2 ratio was controlled using the mixing gases of ($F_2(5\%) + 95\%$ Ar) and pure Ar. AFZO films were deposited on the EagleXG glass and the films' thickness was about 330 nm by controlling the deposition time. After AFZO films were deposited, their crystalline structure and electrical characteristics were well investigated.

[†,§]Corresponding authors.
[*]To cite this article, please refer to its earlier version published in the *International Journal of Modern Physics B*, Volume 34, 2040148 (2020), DOI: 10.1142/S0217979220401487.

Finally, a comparative study of surfaces of AFZO films was investigated using X-ray photoelectron spectroscopy (XPS) to find the variations of the chemical bonds as the deposition temperature and F_2 flow rate were changed.

Keywords: Al and F co-dope; F_2 flow rate; X-ray photoelectron spectroscope; chemical bond.

1. Introduction

The high conductivity (low resistivity) of ZnO-based films can be obtained from the presence of oxygen vacancies and zinc interstitials, and from the addition of different dopants.[1] Al,[2] Ga,[3] F,[4] and Ti[1] are widely used as dopants to enhance the n-type conductivity of ZnO-based films. The doped ZnO films can have conductivity and high transmittance ratio equivalent to indium-doped Tin (ITO) films in the visible light range, but they have higher transmittance ratio in the near infrared range because of their higher plasma frequencies. At the same time, avoiding the over dopant will cause the lattice distortion of deposition films, which will degenerate the films' quality and descend the carriers' mobility, the added dopants are requested to have the similar radii with those of main (ZnO) material. The Zn^{2+} and O^{2-} ions have the radii of 0.74 Å and 1.38 Å and Al^{3+} and F^- ions have the radii of 0.54 Å and 1.31 Å. The radius of Al^{3+} ions is similar to that of Zn^{2+} ions and that of F^- ions is similar to that of O^{2-} ions, for they can be used as the dopant of ZnO films to increase their conductivity.

Recently, many studies had used two different ions as the dopants of the ZnO-based films to investigate their characteristics. Hadri *et al.* used the chemical spray technique to synthesize the In-F co-doped ZnO films on heated glass substrates (350°C).[5] Wang and Chang prepared a ZnO target containing 1 wt% Al_2O_3 and 1.5 wt% ZnF_2 and used it to deposit Al and F co-doped zinc oxide films.[6] In this study, we would investigate the Al and F co-doped ZnO (AFZO) films using Al-doped ZnO ceramic as target and RF sputtering as deposition method, the different flow rates of F_2 were introduced during the deposition process. The effect of different F_2 flow rates on the properties of AFZO films would be well investigated. The chemical structures of AFZO films on different flow rates of F_2 were investigated in full scanning X-ray photoelectron spectroscope (XPS) spectra for clarifying the mechanism of improvement in resistivity.

2. Experimental Procedure

In this work, ZnO (99 wt%, 5 N) and Al_2O_3 (1 wt%, 5 N) was mixed, ground, calcined at 1000°C for 5 h, and sintered at 1350°C to form the

Table 1. Different atmospheres of AZFO films' deposition.

F_2 rate	0%	0.05%	0.1%	0.25%
Ar	20 sccm	20 sccm	20 sccm	20 sccm
$F_2(5\%)$ + 95 % Ar	0 sccm	0.2 sccm	0.4 sccm	1 sccm

Al-doped ZnO (AZO) target with 2 inch in diameter. At first, the chamber pressure was pumped to less than 5×10^{-6} Torr, and then the deposition parameters were controlled at deposition pressure of 5×10^{-3} Torr, deposition power of 100 W, and working distance of 8 cm. Argon was introduced into the chamber with different F_2 flow rates $(F_2/(Ar + F_2) = 0 \sim 0.25\%)$, as Table 1 shows.

EagleXG glass with thickness of 0.7 mm was cleaned using deionized water, acetone, and isopropyl in ultrasonic cleaner. AFZO films' thicknesses were measured using a field emission scanning electron microscopy (FESEM) to measure their depositing rates, and the thickness was controlled at $\sim$340 nm. AFZO films' crystalline structure was identified by X-ray diffraction (XRD). Hall measured equipment could be used to find the resistivity, carrier concentration, and mobility of AFZO films. Optical transmission spectra of all the deposited AFZO films were recorded using a Hitachi U-3300 UV–Vis spectrophotometer in the 300–2000 nm wavelength range. XPS was used to carry out the quantitative and qualitative chemical analyses of AFZO films, which could be used to find chemical bonds and composition of Al, Zn and O elements.

3. Results and Discussion

The XRD patterns of the AFZO films are shown in Fig. 1(a) as a function of F_2 flow rate. As F_2 flow rate was 0%, only the diffraction peaks of the (002) plane and the unapparent (004) plane were observed; as F_2 flow rate was equal to and more than 0.01%, only the diffraction peak of (002) plane was observed. Figure 1(b) shows the electrical measurements of AFZO films. As F_2 flow rate increased, the carrier concentration (n) and mobility (μ) decreased and the resistivity increased. The reasons are that Al^{3+} and F^- ions do not substitute the sites of Zn^{2+} and O^{2-} to form the dopants but they create many defects, which existed in the grain sizes and at the grain boundaries to cause the decreases of n and μ values.[7] As the n value is in the range of 10^{19}–10^{20} cm^{-3}, the scattering of grain boundaries is the main reason to affect μ value; As the n value is in the range of 10^{20}–10^{21} cm^{-3}, the scattering of dopants is the main reason to affect μ value.[8]

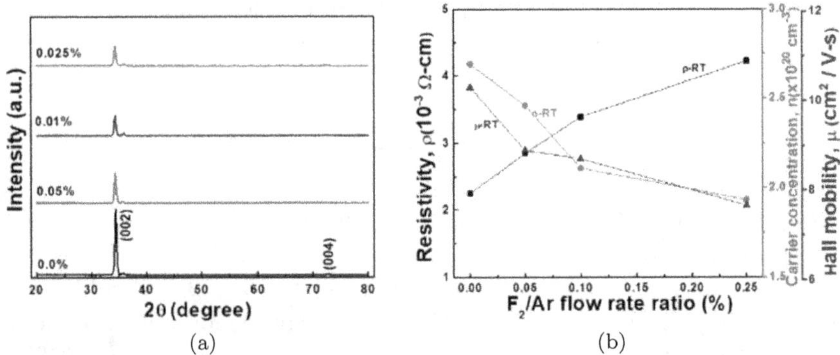

Fig. 1. (Color online) (a) XRD patterns and (b) electrical measurements of AFZO films as a function of F_2 flow rate.

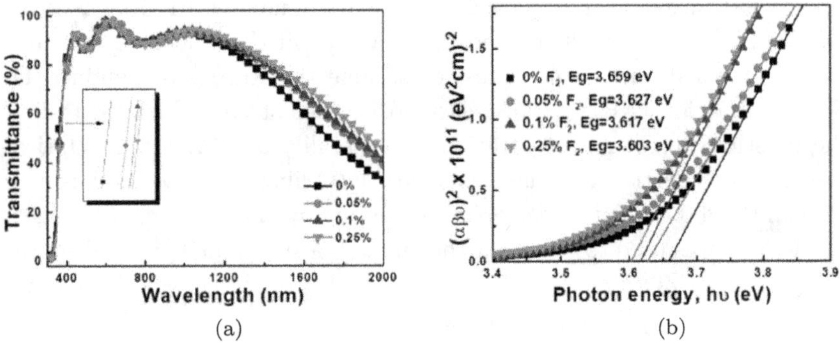

Fig. 2. (Color online) (a) Transmission spectra and (b) $(\alpha h\nu)^2$ versus $h\nu$ (energy) plots of AFZO films as a function of F_2 flow rate.

As Fig. 2(a) shows, when the F_2 flow rate was 0, 0.05%, 0.1% and 0.25%, the average transmittance ratios of AFZO films were 92.1%, 91.7%, 91.9% and 91.7%, respectively. The absorption coefficient (α) and the E_g value of films can be calculated using Eq. (1):

$$T = (1 - R)^2 \exp(-\alpha d) \quad \alpha h\nu = C(h\nu - E_g)^{1/2}, \tag{1}$$

where T is transmittance, R is reflectance, d is film thickness, C is a constant and $h\nu$ is the light energy. Figure 2(b) plots $(\alpha h\nu)^2$ against $h\nu$ (energy) in accordance with Eq. (1), and the E_g values of AFZO films as a function of F_2 flow rate can be obtained from the extrapolated straight line. When the F_2 flow rates were 0, 0.05%, 0.1% and 0.25%, the E_g values of AFZO films were 3.659, 3.627, 3.617 and 3.448 eV. Many factors affect the transmission spectrum and thereby the E_g value, one of the factors is the carrier concentration. When the F_2 flow rate increased from 0% and 0.25%, the n value decreased from 2.69×10^{20} cm^{-3} and 1.93×10^{20} cm^{-3}, for that the E_g value decreased from 3.659 eV and 3.603 eV.

Typical surface O_{1s} peak of AFZO films could be consistently fitted by three nearly Gaussian components, the O_{1s} peak of AFZO films deposited at different F_2 flow rates are shown in Fig. 3. O_{1s} could be divided into three peaks, which were centered at 529.9 ± 0.2, 530.35 ± 0.2 and 532.2 ± 0.2 eV and respective to O_I, O_{II} and O_{III} peaks. The component of O_{1s} spectrum at the binding energy peak of O_I peak is attributed to O–Zn^{2+} bond, the energy peak of O_{II} is thought to be caused by the oxygen vacancy, and the binding energy peak of O_{III} is caused by the chemical absorption on the surfaces of AFZO films.[9] As the F_2 flow rate increased from 0% to 0.25%, the area of O_I (O_{III}) peak increased from 40.88% (20.46%) to 42.86% (21.25%), the area of O_{II} peak decreased from 38.65% to 35.89%.

Typical surface Al_{2p} peak of AFZO films is shown in Fig. 4, where the binding energy peak of 73.0 ± 0.1 eV (Al_I) is contributed by the metal Al, the binding energy peak of 73.57 ± 0.2 eV (Al_{II}) represents the AlO bond, and the binding energy peak of 74.35 ± 0.2 eV (Al_{III}) represents the Al_2O_3 bond.[10] As the F_2 flow rate increased from 0% to 0.25%, the area of Al_I peak decreased from 33.31% to 7.42%, and those of Al_{II} and Al_{III} peaks increased from 41.38% to 49.79% and from 25.31% to 42.79%. The existence of AlO bond (Al_{II}) suggests the existence of the defects in AFZO films and that will cause the decrease of crystallinity, which have been proven in Fig. 1(a).

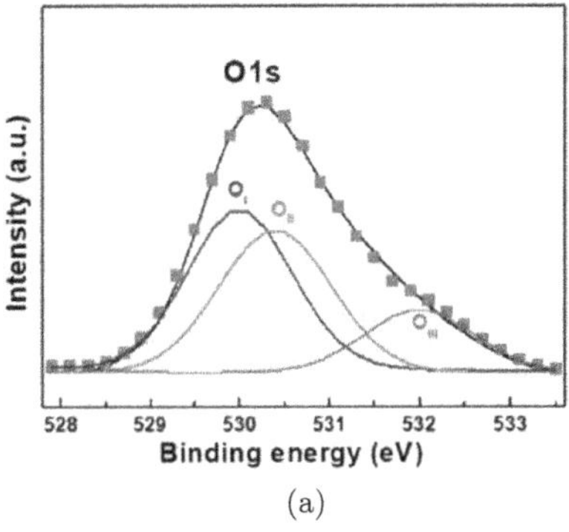

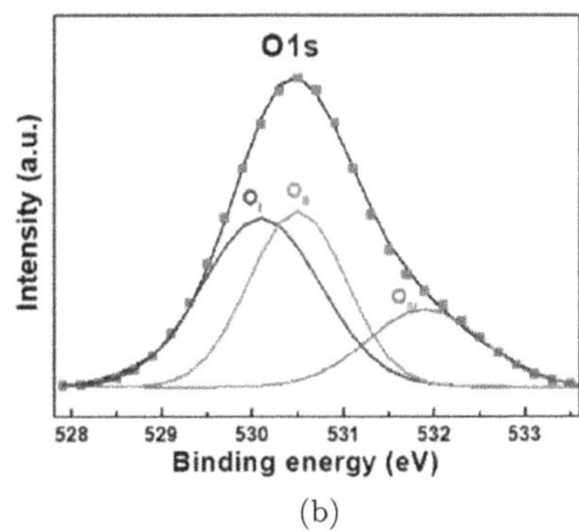

Fig. 3. (Color online) XPS spectra and Gaussian-resolved components of the O_{1s} peak of AFZO films (a) 0% and (b) 0.25% F_2.

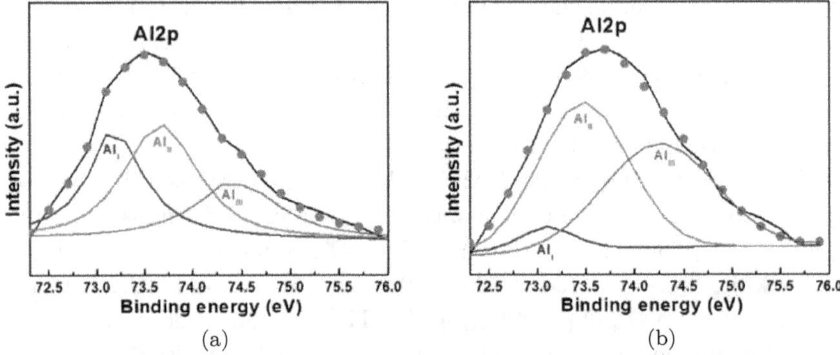

Fig. 4.   (Color online) XPS spectra and Gaussian-resolved components of the Al_{2p} peak of AFZO films (a) 0% and (b) 0.25% F_2.

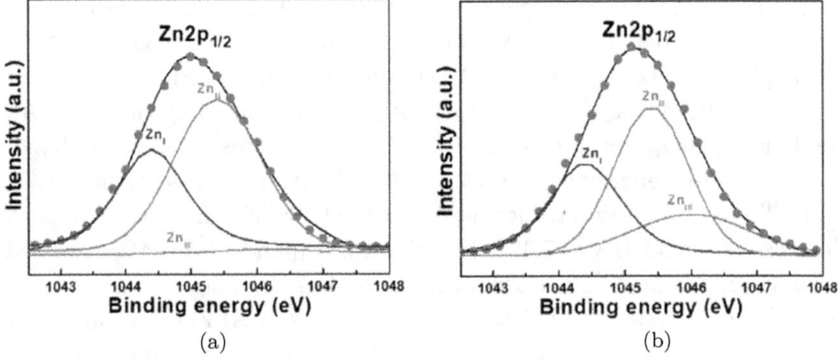

Fig. 5.   (Color online) XPS spectra and Gaussian-resolved components of $Zn_{2p1/2}$ peak of AFZO films (a) 0% and (b) 0.25% F_2.

AFZO films have two fitting peaks located at about 1045 eV and 1022 eV, which are attributed to the $Zn_{2p1/2}$ and $Zn_{2p3/2}$ peaks. Figure 5 shows the surface $Zn_{2p1/2}$ peak of AFZO films, all the $Zn_{2p1/2}$ peak (centered at 1045.2 ± 0.25 eV) of AFZO films could be divided into three peaks, which were centered at 1044.4 ± 0.13, 1045.4 ± 0.16 and 1046.0 ± 0.17 eV and respective to Zn_I, Zn_{II} and Zn_{III} peaks. The Zn_{II} peak is the chemical bond energy of metal Zn, the Zn_{II} peak is bond energy of Zn–O,[11] and the Zn_{III} peak is the chemical bond energy of Zn–F.[12] As the F_2 flow rate increased from 0.00% to 0.25%, the areas of Zn_I and Zn_{II} peaks decreased from 39.95% to 32.89% and from 59.36% to 46.66%, and that of Zn_{III} peak increased from 0.69% to 20.44%. The decrease of the area of Zn_{II} peak is

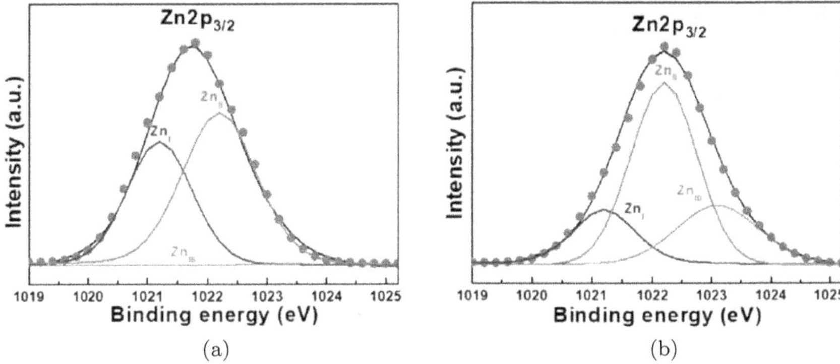

Fig. 6. (Color online) XPS spectra and Gaussian-resolved components of $Zn_{2p3/2}$ peak of AFZO films (a) 0% F_2 and (b) 0.25% F_2.

caused by Al^{3+} ions occupying the sites of Zn^{2+} ions, the number of Zn–O bond decreases, and the results are consistent with the variations of the Al_{2p} peak.

The surface $Zn_{2p3/2}$ peaks (centered at 1022.1 ± 0.14 eV) of AFZO films deposited at different F_2 flow rates and temperatures are shown in Fig. 6. The low energy bond of 1021.2 ± 0.12 eV is the chemical energy bond of the Zn metal, the high-energy bond of 1022.2 ± 0.11 eV is the chemical energy bond of the Zn–O[11] and 1023.1 ± 0.12 eV is the chemical bond energy of Zn–F.[12] As the d F_2 flow rate reached 0%, the area of Zn_I peak decreased from 45.47% to 18.46%, and those of Zn_{II} and Zn_{III} peaks increased from 54.01% to 57.37% and from 0.52% to 24.17%.

4. Conclusions

As the F_2 flow rate increased from 0% to 0.25%, the average transmittance ratio of AFZO films was in the range of 92.1% and 91.7% and the E_g value of AFZO films decreased from 3.659 eV to 3.448 eV. The decrease of n value from 2.69×10^{20} cm^{-3} and 1.93×10^{20} cm^{-3} is the reason to cause the variation of E_g value. As F_2 flow rate increased, Al^{3+} and F^- ions would create many defects, which existed in the grain sizes and at the grain boundaries to cause the decrease in n and μ values. For the chemical bonds, as the F_2 flow rate increased from 0% to 0.25%, the areas of O_I, O_{III}, Al_{II}, Al_{III}, Zn_{III} of $Zn_{2p1/2}$ and Zn_{II} and Zn_{II} of $Zn_{2p3/2}$ peaks increased and those of O_{II}, Al_I, Zn_I and Zn_{II} of $Zn_{2p1/2}$ and Zn_I of $Zn_{2p3/2}$ peaks decreased.

Acknowledgments

This work was supported by projects under Ministry of Science and Technology Grant Nos. MOST 108-2221-E-390-005 and MOST 108-2622-E-390-002-CC3.

References

1. C. C. Huang *et al.*, *Nanoscale Res. Lett.* **8**, 206 (2013).
2. F. H. Wang *et al.*, *J. Nanomaterials* **2014**, 857614 (2014).
3. F. H. Wang *et al.*, *Nanomaterials* **6**, 88 (2016).
4. Y. Du *et al.*, *Mod. Phys. Lett. B* **33**, 1940040 (2019).
5. A. Hadri *et al.*, *Thin Solid Films* **601**, 7 (2016).
6. F. S. Wang and C. L. Chang, *Appl. Surf. Sci.* **370**, 83 (2016).
7. T. C. Lin *et al.*, *Appl. Surf. Sci.* **258**, 3302 (2012).
8. T. Minami *et al.*, *J. Cryst. Growth* **117**, 370 (1992).
9. D. K. Kim and H. B. Kim, *J. Alloys Compd.* **522**, 69 (2012).
10. J. P. Kar *et al.*, *Solid State Electron.* **54**, 1447 (2010).
11. F. H. Wang *et al.*, *Surf. Coatings Technol.* **205**, 5269 (2011).
12. R. Mayer, *The Artist's Handbook of Materials and Techniques*, 5th edn. (Faber, 1991), p. 784.

The structural correlation and mechanical properties in amorphous hafnium oxide under pressure*

Nguyen-Hoang Thoan, Nguyen-Trung Do and Nguyen-Ngoc Trung[†]

School of Engineering Physics,
Hanoi University of Science and Technology,
No. 1, Dai Co Viet Road, Hanoi 100000, Vietnam
[†]*trung.nguyenngoc@hust.edu.vn*

Le-Van Vinh[‡]

Phenikaa University, Yen Nghia,
Ha Dong, Hanoi 100000, Vietnam
vinh.levan@phenikaa-uni.edu.vn

The classical molecular dynamics (MD) technique was used to investigate the atomistic structure of amorphous hafnium oxide (HfO_2) under pressure. The local atomic structure and the liquid-solid transition of HfO_2 were analyzed for the pair radical distribution functions, bond angle distributions and coordination number. The simulation reveals that although the fractions of structural units HfO_x and OHf_y strongly change with the density, the partial bond angle distributions of these structural units are almost identical for all constructed models. This result has enabled us to establish a relationship between the bond angle distributions and the fractions of structural units.

Keywords: Amorphous hafnia; molecular dynamic; atomic structure; pressure.

1. Introduction

Hafnium oxide HfO_2 has been extensively studied for use as silicon dioxide replacements in gate insulating layers in CMOS devices and resistive RAM devices. Solid HfO_2 can be formed at different phases including amorphous, monoclinic, tetragonal and cubic. However, HfO_2 often deposits as

---

[†,‡]Corresponding authors.

*To cite this article, please refer to its earlier version published in the *International Journal of Modern Physics B*, Volume 34, 2040149 (2020), DOI: 10.1142/S0217979220401499.

an amorphous phase by CVD or ALD fabrication processes. Many comprehensive studies of the structure and thermodynamics of the HfO_2 crystal and amorphous phases have been conducted intensively by both experiments and simulations in recent years.[1–3] Recently, we investigated molecular dynamics (MD) simulation results on liquid-amorphous transition of HfO_2 under cooling process.[4] Tapily *et al.* demonstrated that the initial low-temperature ALD deposited HfO_2 in its amorphous state is harder and stiffer than the polycrystalline form of HfO_2.[5] Therefore, this has motivated us to carry out this study. We will report on the results of density and structural evolution of a-HfO_2 under compression and discuss the unit structure HfO_x and mechanical behavior of a-HfO_2.

2. Computational Procedures

The MD simulation has been performed using an initial configuration containing 2000 Hf atoms and 4000 oxygen ones in a simulation box corresponding to the density of 10.2 g $\cdot$ cm^{-3} with a condition that no two atoms be closer than 1.0 Å. We apply the Morse-KBS combined potential as mentioned in our previous study, Ref. 4. The long-range Coulomb interactions were calculated with the Ewald summation technique, which is applied to three-dimensional periodic boundary conditions. The Verlet algorithm with a time step of 1.5 fs is adopted and the simulation was executed at a constant pressure. The scaling method was adopted during the simulation to control the system temperature. The initial configuration was equilibrated for 50,000 static relaxation steps, and for the next 50,000 steps at 5000 K by MD method. Sequentially, the samples were cooled down to 300 K within 470,000 steps without or with applying a pressure during cooling process ($P = 0 - 100$ GPa). To determine the coordination number (CN), bond angle distribution (BAD), the structural unit HfO_x and OHf_y linkages, we used the cutoff distance $R_{c(\text{Hf-O})} = 2.57$ Å, which is a cutoff radius as the first minimum after peak of Hf–O pair radical distribution functions (RDF). Strain-stress simulations with a strain rate of 2×10^{11} s^{-1} were calculated following the computational procedures which were described in detail elsewhere.[6]

3. Results and Discussion

At 300 K, HfO_2 density increases gradually from 9.23 to 12.26 g $\cdot$ cm^{-3} with increasing pressure from 0 GPa to 100 GPa. Figure 1 presents the partial and total RDF $g(r)$ of a-HfO_2 samples at different pressures. For the total RDFs $g_N(r)$ exhibiting the short-order structure of amorphous

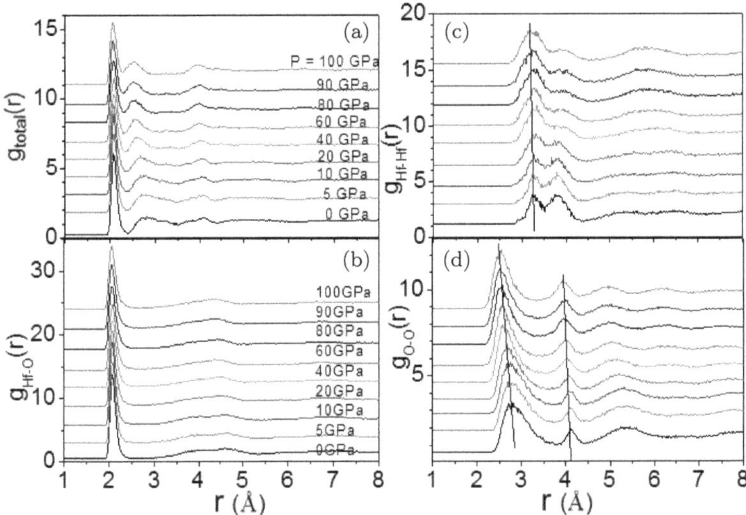

Fig. 1. (Color online) The total (a) and partial (b)–(d) RDFs of a-HfO$_2$ with different pressures. The straight lines are to guide the eyes.

phase [Fig. 1(a)], the first peak is almost unchanged, while the second peak shifts to the left from 2.83 Å to 2.55 Å with increasing pressure. At 0 GPa (no pressure), the $g_N(r)$ of the model is in good agreement with the experimental data.[7] The first peak of the $g_N(r)$ is contributed from the pair RDF $g_{Hf\text{-}O}(r)$ and the second and third peaks are contributed from the pair RDFs $g_{O\text{-}O}(r)$ and $g_{Hf\text{-}Hf}(r)$. All $g_{Hf\text{-}O}(r)$ functions of all samples exhibit a solid state with disorder structure (amorphous phase), indicating that no pressure-induced phase transformation occurs in a-HfO$_2$ systems. For the $g_{Hf\text{-}O}(r)$, Fig. 1(b), the shape and peak position are almost unchanged except the height of the first peak with increasing pressure. Hf–O bond distance is almost unchanged at $r = 2.08 \pm 0.02$ Å, which is shorter than the average value of 2.14 Å in the monoclinic crystal.[2] The $g_{Hf\text{-}Hf}(r)$ in Fig. 1(c) present a dual peak between 3 Å and 4 Å. The first one slightly shifts to the left from 3.26 Å to 3.20 Å; while the second one at ~3.80 Å decreases its intensity with increasing pressure. The first peak of the $g_{O\text{-}O}(r)$, Fig. 1(d), also shifts to the left from 2.77 Å to 2.51 Å upon compression, and the second from 4.12 Å to 3.94 Å. These results are close to that in previous simulation and experimental studies.[3,7–9]

The network structure of a-HfO$_2$ samples consists of HfO$_x$ units and OHf$_y$ linkages as presented in Fig. 2. One can see that HfO$_6$ units and OHf$_3$ linkages are dominant for samples at 0 GPa. However, the increasing

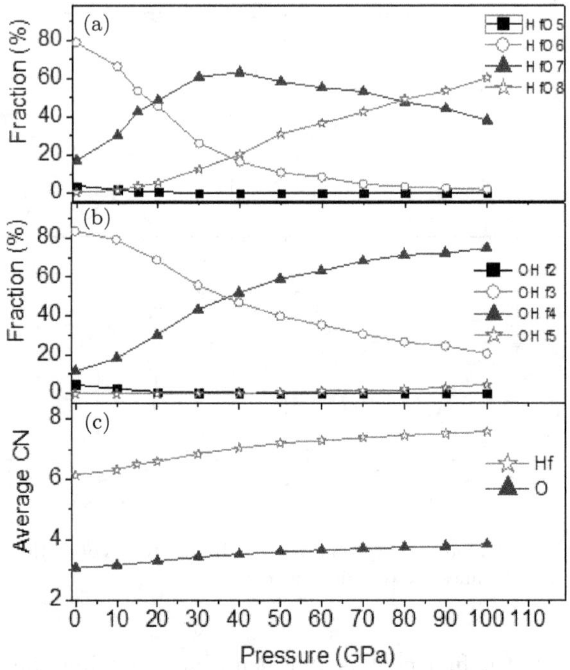

Fig. 2.   (Color online) Fractions of (a) HfO_x units; (b) OHf_y linkages and (c) the average CN of Hf and O.

of pressure induces in a-HfO_2 a change in the Hf CN, which passes, on average, from 6 to 8. Fraction of HfO_6 units decreases drastically from 78.7% at $P = 0$ GPa to 1.77% at $P = 100$ GPa. While, fraction of HfO_8 units increases drastically from 0.35% at $P = 0$ GPa to 60.2% at $P = 100$ GPa. On average, the Hf–O CN ranges from 6.14 to 7.58 as shown in Fig. 2(c). Figure 3 displays snapshots of HfO_5 and HfO_8 units in three samples at $P = 10, 20, 40$ GPa, respectively, showing the same type of units to form clusters via common oxygen.

The same trend has been observed also for the average oxygen CN. As shown in Fig. 2(b), fraction of OHf_3 units decreases continuously from 63.6% at $P = 0$ GPa to 20.4% at $P = 100$ GPa. While, fraction of OHf_4 units increases continuously from 11.6% at $P = 0$ GPa to 74.8% at $P = 100$ GPa. On average, the O–Hf CN ranges from 3.07 to 3.84 as shown in Fig. 2(c). Cressoli *et al.* simulated amorphous HfO_2 and also found a prevalence of 6- and 7-fold coordinated Hf and 3- and 4-fold coordinated oxygen atoms.[10] However, there is a large variation in the predicted packing

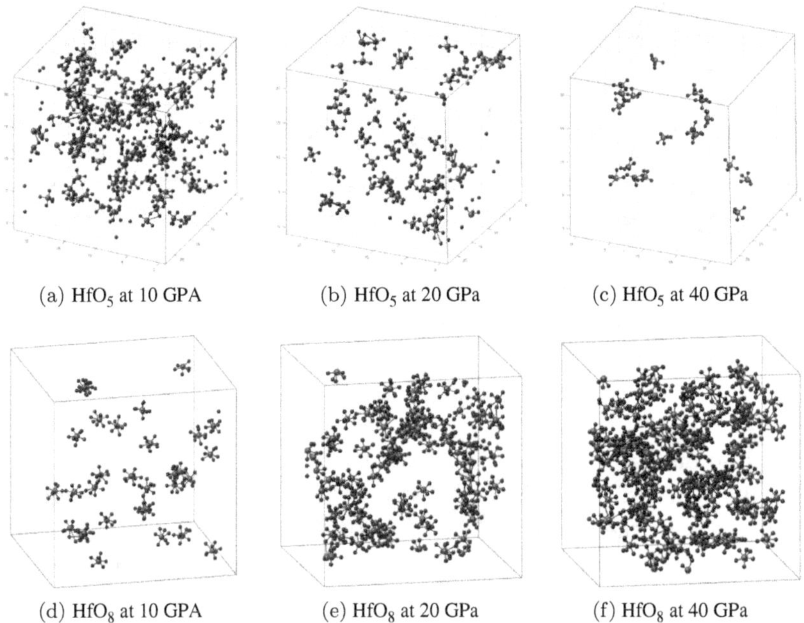

(a) HfO$_5$ at 10 GPA   (b) HfO$_5$ at 20 GPa   (c) HfO$_5$ at 40 GPa

(d) HfO$_8$ at 10 GPA   (e) HfO$_8$ at 20 GPa   (f) HfO$_8$ at 40 GPa

Fig. 3. (Color online) Visually atomic clusters of HfO$_5$ units (a)–(c) and HfO$_8$ units (d)–(f) in samples under compression at $P = 10$, 20 and 40 GPa. Red: Hf; and Blue: O.

between the polyhedra in the different models. A study combining both classical MD and DFT calculations found that the domination of 6-fold Hf atoms and 3-fold O atoms led to significantly more corner-sharing polyhedral than edge sharing.[3] According to Gallington *et al.*, corner and edge-shared HfO$_7$ polyhedra present in the monoclinic phase compared to the solely edge-shared HfO$_8$ polyhedra that comprise the cubic phase.[7] In the monoclinic phase, Hf atoms are surrounded by 7 oxygen atoms, with an asymmetric arrangement of Hf–O bond distances in the range 2.03–2.25 Å, corresponding to a mixture of 7-edge shared polyhedra and 4-corner shared polyhedra in the unit cell.[11] In monoclinic HfO$_2$, 50% of the O atoms have a CN equal to 3% and 50% have a CN equal to 4. Gallington found essentially the same numbers are observed both in the melt (O–Hf CN = 3.5) and the amorphous solid (O–Hf CN = 3.4); likewise, the distribution of bond lengths is highly asymmetric, with ∼30% of the atoms involved in extended O–Hf bonds.[7] This behavior might be associated with a more compact structure, which is promoted by higher density, being related to the smaller volume available for the system to relax.

More details about the local atomic structures can be inferred from BAD: O–Hf–O BAD describes the topology inside the structural units, and the Hf–O–Hf one gives the connectivity between them. As shown in Fig. 4(a), both the peaks of O–Hf–O BADs shift to the left upon compression upto 40 GPa: the main one shifts from 78° to 73°, and the other from 160° to ~136°. In addition, the height of both two peaks increases with increasing pressure upto 40 GPa. These results may relate to changing of main structure units from HfO_6 to HfO_7 when increasing pressure. When increasing pressure from 40 to 100 GPa, the position of both two peaks and the height of the main peak are almost unchanged, while the height of the second minor peak increases slightly. Broglia *et al.* also showed that O–Hf–O angles are insensitive to density, and the Hf–O CN is most affected by density.[12] In this study, changing of pressure results in change of sample's density, so our results agree with Broglia's.[12]

Figure 4(b) presents Hf–O–Hf BADs of samples with different pressures. Compared with the results in Fig. 2(b), the Hf–O–Hf BAD at $P = 0$ GPa is mainly contributed from the Hf–O–Hf bond angles of OHf_3 linkages, at which a main peak locates at 99°, and a small peak at 126°. These bond angles have been also observed in Ref. 12 with a difference of 2°. Upon increasing pressure, the peak's positions shift slightly to the right, the intensity of the first peak slightly reduces, and the second peak's intensity

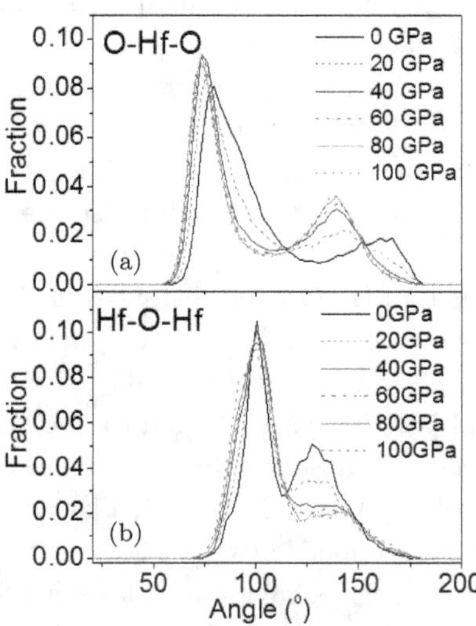

Fig. 4. (Color online) The (a) O–Hf–O and (b) Hf–O–Hf BADs for samples with different pressures.

reduces more. At $P = 100$ GPa, the Hf–O–Hf BAD is mainly contributed from the Hf–O–Hf bond angles of the OHf$_4$ linkages.

Our recent study, Ref. 4, reported that the almost corner-sharing bond appears in the HfO$_5$–HfO$_5$ connectivity whereas three types (corner-, edge- and face-) of sharing bonds appear in the HfO$_6$–HfO$_6$ and HfO$_7$–HfO$_7$ connections. In this study, we also found that three types, corner-, edge- and face-, of sharing bonds appear in the HfO$_8$–HfO$_8$ connections. For HfO$_5$–HfO$_6$ and HfO$_5$–HfO$_7$ connection, corner-sharing bonds account for dominant bonds whereas edge-sharing bonds account for minor bonds and face-sharing bonds do not appear. For HfO$_5$–HfO$_7$ connections, edge- and face-sharing bonds are the predominant part whereas small amounts of face-sharing bonds appear. The strong shift of main structure units from HfO$_5$ and HfO$_6$ to HfO$_7$ and HfO$_8$ upon compression results in deduce of HfO$_5$- and HfO$_6$-related connectivities and increase of HfO$_7$- and HfO$_8$-related connectivities. These also affect the BADs, because each connectivity has its own bond angle. All evidences indicate that the same type of HfO$_x$ units connects together to create the same HfO$_x$ cluster with common oxygen bridges.

We also investigate the mechanical behavior of a-HfO$_2$ upon compression. Figure 5 presents stress-strain curves for a-HfO$_2$ samples upon compression. Young's modulus is estimated by the slope of the stress-strain

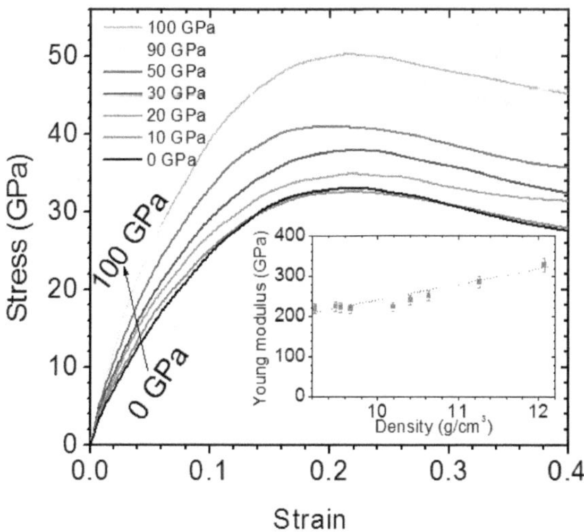

Fig. 5.   (Color online) Stress-strain curves for a-HfO$_2$ samples upon compression. The insert: Young's modulus of a-HfO$_2$ systems as a function of a-HfO$_2$ density.

curves in the linear regions. The stress continues to increase after the yield point reaches a plateau and then decreases. The calculated Young's modulus is comparable with experimental one for amorphous HfO_2.[5,13] Young's modulus increases with increasing density, as shown in the insert, due to the increasing fraction of high CN.

4. Conclusion

MD has been used to investigate the atomic structural characteristics and mechanical behavior of a-HfO_2 system upon compression. Under pressure, the transition of HfO_x units and OHf_y linkages occurs from low to higher CNs. The same type of units tends to aggregate into clusters with common oxygen bridges. The Young modulus increases upon compression due to the increasing fraction of high CNs.

Acknowledgment

This research is funded by Hanoi University of Science and Technology (HUST) under project number T2020-PC-058.

References

1. A. A. Demkov and A. Navrotsky, *Materials Fundamentals of Gate Dielectrics* (Springer, Netherlands, 2005).
2. W. L. Scopel, A. J. R. da Silva and A. Fazzio, *Phys. Rev. B* **77**, 172101 (2008).
3. Y. Wang *et al.*, *Phys. Rev. B* **85**, 224110 (2012).
4. T. H. Nguyen, V. V. Le and T. N. Nguyen, *Vacuum* **161**, 251 (2019).
5. K. Tapily *et al.*, *Int. J. Surf. Sci. Eng.* **5**, 193 (2011).
6. V.-V. Le *et al.*, *Comput. Mater. Sci.* **79**, 110 (2013).
7. L. C. Gallington *et al.*, *Materials (Basel)* **10**, 1290 (2017).
8. T. J. Chen and C. L. Kuo, *J. Appl. Phys.* **110**, 064105 (2011).
9. J. P. Trinastic *et al.*, *J. Chem. Phys.* **139**, 154506 (2013).
10. D. Ceresoli and D. Vanderbilt, *Phys. Rev. B* **74**, 5 (2006).
11. C. Wang, M. Zinkevich and F. Aldinger, *J. Am. Ceram. Soc.* **89**, 3751 (2006).
12. G. Broglia, G. Ori, L. Larcher and M. Montorsi, *Model. Simul. Mater. Sci. Eng.* **22**, 065006 (2014).
13. M. Berdova *et al.*, *J. Vac. Sci. Technol. A* **34**, 051510 (2016).

https://doi.org/10.1142/9789819807727_0012

Vehicle speed estimation using two roadside passive infrared sensors*

Van Quang Vu, Van Linh Ngo and Toan Thang Vu[†]

*Hanoi University of Science and Technology,
Hanoi 100000, Vietnam*
[†]*thang.vutoan@hust.edu.vn*

Van Phuc Doan

Institute of Chemistry and Material, Hanoi 122100, Vietnam

This paper presents a method for estimating vehicle speed using two roadside passive infrared (PIR) sensors whose optical axes are parallel to each other and perpendicular to the moving direction of vehicles. The vehicle speed was calculated based on the time lag between the two signals received by the PIR sensors, which was evaluated by using cross-correlation analysis. The experiments show that the method has an error of less than 5 km/h over a speed range of 20–60 km/h.

Keywords: Vehicle speed; passive infrared (PIR) sensor; cross-correlation.

1. Introduction

In Intelligent Transportation Systems (ITS), using passive infrared (PIR) sensors is one of the nonintrusive technologies for vehicle detection and surveillance in which the sensors or transducers are not installed directly onto or into the road surface but are mounted overhead or on the roadside.[1] Others in this group include video image processing, microwave radar, ultrasonic sensors, etc. Such technologies have their own advantages as well as disadvantages, as mentioned in Refs. 1 and 2. PIR sensors, despite their

---

[†]Corresponding author.
*To cite this article, please refer to its earlier version published in the *International Journal of Modern Physics B*, Volume 34, 2040151 (2020), DOI: 10.1142/S0217979220401517.

issues when working in hazard weather conditions, may be a good solution for cost-effective and low power ITS systems.

In many applications, PIR sensors are employed for tracking and localizing heat sources, such as humans, animals, or vehicles, because the sensors can detect the infrared energy emitted from such sources.[3–5] However, studies on vehicle speed estimation using PIR sensors are still limited. Among them, in Ref. 2, a traffic sensor node consisting of an ultrasonic ranger and three PIR sensors was mounted overhead to estimate the vehicle speed and length, and to classify the kinds of vehicles. However, in such outdoor applications, the sensors require considerable work in terms of installation, and the measurement accuracy faces the issue of a low sampling rate. In this paper, a portable roadside system that uses two PIR sensors is presented for estimating the speed of different kinds of vehicles. With the assumption that the sensors' optical axes are parallel to each other, this system is able to measure the speed of vehicles that move at different distances from the sensors.

2. Measuring System

The system uses two PIR thermopile sensors, which convert the difference between the temperature of the sensor surface and ambient temperature into voltage.[6] When the vehicle moves into the field of view (FOV) of the sensor, as its engine is working, it emits infrared energy that partially transmits to the sensor. As a result, the sensor surface temperature becomes hotter than the ambient temperature; thus, an output positive voltage is established. To concentrate the infrared energy onto the sensor surface and to narrow the FOV of the sensor, germanium lenses have been used. The lenses and sensors work well with far infrared wavelengths of the range of 8–14 μm, which is suitable for most objects with surface temperatures from $-60°$C to 90°C according to the Wein displacement law.[7] The housings for the sensor and lens are 3D printed and fixed in clamps on a flat aluminum table (Fig. 1) so that the optical axes of the two sensors are parallel to each other. Consequently, the sensor output signals are amplified with a gain of 60 dB and then sampled by a built-in 12-bit analog-to-digital converter of a STM32L432KC microcontroller with a sampling rate of 1000 Hz. The sampling rate is important for the accuracy and resolution of speed estimation determined via cross-correlation analysis.

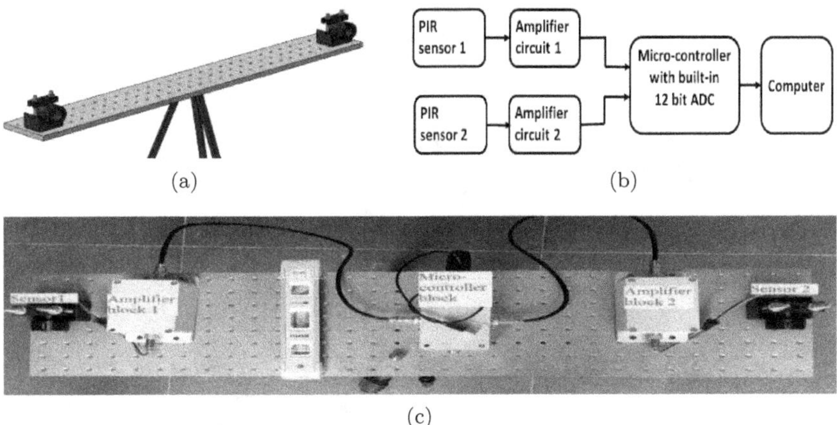

Fig. 1. (Color online) Measuring system: (a) 3D design; (b) functional diagram and (c) top view of the real system.

3. Experiments

The speed measurement system was placed on the roadside so that the sensor optical axes were perpendicular to the moving direction of the considered vehicles (Fig. 2). A video camera with a frame rate of 200 fps was used to evaluate the vehicle speed from reference points by cutting the video into frames to track the vehicle position.

Targeted vehicles with different lengths and shapes move with the speed of upto 60 km/h, as it is limited inside the particular residential community. The distance from the moving lane of the vehicle to the measurement system varied from 6 m to 19.5 m.

4. Signal Processing and Speed Estimation Method

Before analyzing and using the algorithm for signal processing and speed estimation, the following assumptions were made for these experiments:

(i) The ambient temperature did not change during the measurement.
(ii) The targeted vehicle moves at a constant speed when it passed across the two roadside sensors owing to the small placement distance of the two sensors and their narrow FOVs. With this assumption, the shapes of the output signals of the two sensors for one targeted vehicle will

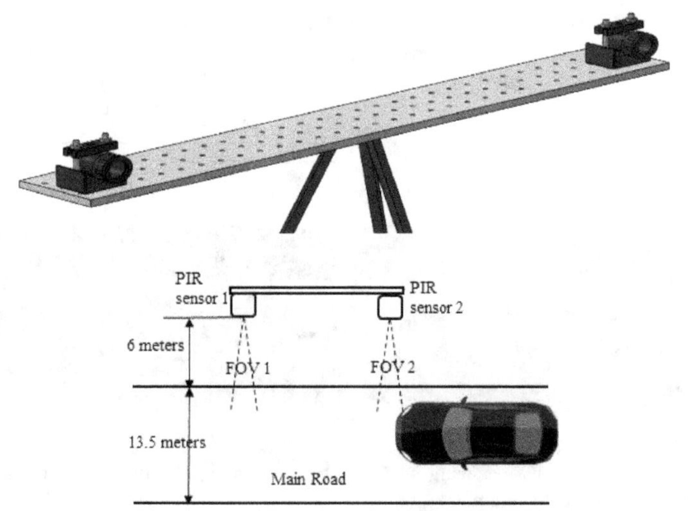

Fig. 2.   Illustration of the experimental setup.

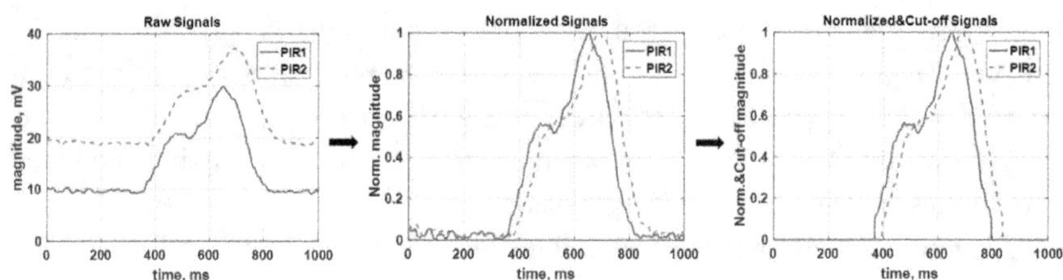

Fig. 3.   (Color online) Signal processing stages.

be similar to each other. This is the basis for estimating vehicle speed using cross-correlation analysis.

During the signal processing stages, the DC offset of the signal was removed, and then the signal magnitude was normalized. A threshold of 0.1 was practically applied to the normalized signals to indicate the presence of a vehicle (Fig. 3).

The vehicle speed was calculated as

$$v = \frac{d}{\Delta t}. \tag{1}$$

Here, d is the distance between the two sensors, which was set to 0.6 m in this study; Δt is the time delay of samples of the two output processed

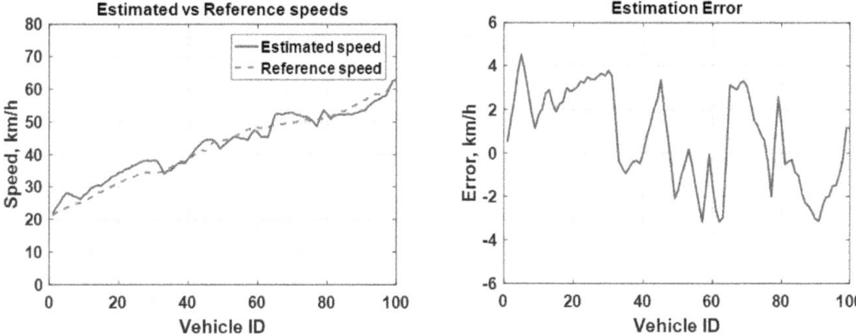

Fig. 4.   (Color online) Vehicle speed estimation results.

signals, which can be defined as

$$\Delta t = t_s \cdot \arg\max_n F(n), \tag{2}$$

$$F(n) = \sum_{k=0}^{N-1} S_1(k) \cdot S_2(k-n), \quad -(N-1) < n < (N-1), \tag{3}$$

where t_s is the sampling period, S_1 and S_2 are the magnitudes of the processed signals, $F(n)$ is their cross-correlation function, and N is the number of samples. In practice, when the number of samples is high, an invert fast Fourier transformation (IFFT) can be used for performing cross-correlation at a fast speed.[8]

An analysis based on Eqs. (1)–(3) was applied to the processed signals to estimate the speed of targeted vehicles. Figure 4 shows a comparison between the speed estimation result using this method and the reference speed using a video camera. The result shows an error of less than 5 km/h in the speed range of 20–60 km/h. Some factors that affect the accuracy may include the following: background noise interferences, such as other heat sources that appear during the measurement, the predefined assumptions are not consistent, and the optical axes of the two sensors are not perfectly parallel to each other.

5. Conclusion

The roadside system presented in this paper provides an acceptable result for estimating speeds of different types of vehicles. The system uses two PIR sensors for signal acquisition and cross-correlation analysis was performed for calculating vehicle speed. This system is portable and easily installed for measurements. In future work, we will focus on developing methods and algorithms that allow the system to work under conditions that are more dynamic.

Acknowledgment

The authors gratefully acknowledge the financial support of the Project of Hanoi University of Science and Technology with code T2018-T-004.

References

1. L. E. Y. Mimbela *et al.*, Technical Report (FHWA, 2000), pp. 183–187.
2. E. Odat *et al.*, *IEEE Trans. Intell. Transp. Syst.* **19**, 1593 (2018).
3. J. Yun and S.-S. Lee, *Sensors* **14**, 8057 (2014).
4. P. Zappi *et al.*, *IEEE Sensors J.* **10**, 1486 (2010).
5. T. M. Hussain *et al.*, *IEEE Trans. Veh. Technol.* **44**, 683 (1995).
6. J. Fraden, *Handbook of Modern Sensors: Physics, Designs, and Applications*, 4th edn. (Springer, 2000), pp. 483–487.
7. J. S. Accetta, *The Infrared and Electro-Optical Systems Handbook, Volume 1: Sources of Radiation*, 1st edn. (Society of Photo Optical, 1993), pp. 11–14.
8. S. Taghvaeeyan and R. Rajamani, *IEEE Trans. Intell. Transp. Syst.* **15**, 73 (2014).

Effects of removing temperature of reduction gas on the luminescence characteristics of Li_2BaSiO_4:0.003EU^{2+} green phosphor*

Jie Li[†] and Yang-Ming Lu[‡]

*Department of Electrical Engineering, National University of Tainan,
Tainan City 700, Taiwan*
[†] *200461000055@jmu.edu.cn*
[‡] *ymlu@mail.nutn.edu.tw*

Qing-Hao Yang[§] and Cheng-Fu Yang[¶]

*Department of Chemical and Materials Engineering,
National University of Kaohsiung, Kaohsiung 811, Taiwan*
[§] *a1054304@mail.nuk.edu.tw*
[¶] *cfyang@nuk.edu.tw*

We had successfully synthesized green-emitting phosphors based on Li_2BaSiO_4 material activated by bivalent europium ions (Eu^{2+}) using a solid-state reaction method in a reducing gas environment and investigated their luminescence properties. The Li_2BaSiO_4:0.003Eu^{2+} (LSB-Eu) phosphors were synthesized at 850°C for 1 h, and the reduction gas was removed at 500°C, 600°C, 700°C and 800°C, respectively. XRD pattern showed that the Li_2BaSiO_4, Ba_2SiO_4 and Li_4SiO_4 phases were observed in the synthesized Li_2BaSiO_4 composition. As the reduction gas was removed at 800°C, the LSB-Eu phosphor emitted a weak red light rather than a green light. Two weak emission peaks were found at about 588 nm and 613 nm corresponding to $^5D_0 \rightarrow {}^7F_1$ and $^5D_0 \rightarrow {}^7F_2$ transitions. As temperature to remove the reduction gas was lower than 800°C, the emission spectra of LSB-Eu phosphors reveled a broad peak centered at 501 nm, which emitted a green color. The intensity of photoluminescence excitation (PLE) photoluminescence emission (PL) spectra increased as the removing temperature was decreased from 700°C to 500°C and saturated

---

[¶]Corresponding author.
*To cite this article, please refer to its earlier version published in the *International Journal of Modern Physics B*, Volume 34, 2040159 (2020), DOI: 10.1142/S0217979220401591.

at 500°C. These results show that LSB-Eu can be a noteworthy candidate of green-emitting phosphor for the investigation of white light-emitting diodes (WLEDs).

Keywords: Removing temperature; reduction gas; Li_2BaSiO_4:0.003Eu^{2+}; green phosphor.

1. Introduction

The adopted compositions of host materials and the adopted activators are two critical factors to affect the emission light of investigated phosphors. In order to investigate the phosphors with suitable emission color, many researchers still keep exploring luminescent properties for different compositions of host materials with different activators. In the past, $Li_2(Ca, Sr, Ba)SiO_4$ had been adopted as host materials and different ions had been adopted as activators to investigate the phosphors with different colors. Li *et al.* used $Li_2Ca_{0.4}Sr_{0.6}SiO_4$ as host material and Tb^{3+} as activator to investigate green phosphor.[1] Kulshreshtha *et al.* had investigated the ternary compositions of Li_2CaSiO_4:Eu^{2+}-Li_2BaSiO_4:Eu^{2+}-Li_2SrSiO_4:Eu^{2+}, which would emit bluish green, pure green, green yellow and yellow luminescence.[2] Some studies also found that the specific binary composition of $Li_2(Ba_{1-x}Sr_x)SiO_4$:Eu^{2+} (0.28 < x < 0.56) showed the best performance of green emission.[2,3] Feng *et al.* found that in the $Li_2(Ba_{1-x}Sr_{x-0.003})SiO_4$:0.003$Eu^{2+}$ compositions, the $Li_2(Ba_{0.6}Sr_{0.397})SiO_4$:0.003$Eu^{2+}$ would have optimal emission intensity.[4] When rare earth elements are used to dope phosphor-based materials, they can play essential roles in producing different emission wavelengths and enhancing phosphors having excellent luminescence properties.

Europium (Eu) is the most used activator because it has the two ions' types, trivalent (Eu^{3+}) and bivalent (Eu^{2+}). They can play an important role in phosphors because of the abundant and different emission colors resulting from the d-f (Eu^{2+}) or f-f (Eu^{3+}) transitions.[5] In 4f–4f transitions, Eu^{3+} is a red-emitting activator because of its $^5D_0 \rightarrow {}^7F_c$ transitions ($c = 0, 1, 2$ and 3).[6] Materials activated by Eu^{3+} ions can produce sharp emission lines in the orange, orange-red, red regions because they have $^5D_0 \rightarrow {}^7F_c$ transitions, and they are considered as red sources.[7] Eu^{2+} is a blue-[8] or green-emitting activators[9] (which is mainly dependent on the adopted host materials) because of its transition from different excited states, for example, the higher 5d levels of 5d-4f transitions to the ground state. In an unexpected research, we had found that as the reduction gas of Li_2BaSiO_4:0.003Eu^{2+} (LSB-Eu) composition was removed at 800°C, it could not act as green phosphor and it would emit weak red light. In the past, a literature survey reveals no publications on how removing temperature of reduction gas (RTRS) affects the luminescence properties of LSB-Eu green phosphor. For that, we had well investigated the LSB-Eu phosphors by solid-state reaction method in reduction atmosphere and we would show the important novelty in the synthesized LSB-Eu phosphors.

2. Experimental Procedure

First, $SrCO_3$ (99.9%), SiO_2(99.9%), Li_2CO_3 (99.9%) and Eu_2O_3 (99.9%) were as the raw materials, then they were mixed to formulate Li_2BaSiO_4:0.003Eu^{2+} (LSB-Eu) powder and ball-milled for 2 h with absolute alcohol. After being dried and ground, the powders were synthesized at 850°C via a solid-state reaction technique in a reduction atmosphere of 96% $N_2 + 4\%$ H_2 for 1 h. The removing temperatures of reduction gas (RTRS) were 500°C, 600°C, 700°C and 800°C, and the samples were abbreviated as LSB-Eu-500, LSB-Eu-600, LSB-Eu-700 and LSB-Eu-800 phosphor, respectively. The crystalline structures of the synthesized LSB-Eu powders were measured using X-ray diffraction (XRD) patterns with Cu Kα radiation ($\lambda = 1.5418$ Å) and at a scanning speed of 2° per minute. Scanning electron microscopy was used to observe the morphology of LSB-Eu powder with an accelerating voltage of 10 kV. Photoluminescence excitation spectra (PLE) of LSB-Eu phosphors recorded in the wavelength range of 200–450 nm, equipped with an emitting light at a maximum emission wavelength λ_{max} of 501 nm for LSB-Eu phosphors, and the photoluminescence (PL) properties were also recorded at room temperature in the wavelength range of 400–700 nm, equipped with an emitting light at an excitation wavelength λ_{ex} of 283 nm or 363 nm for LSB-Eu phosphors. Both the PLE and PL spectra were measured on a Hitachi F-4500 fluorescence spectrophotometer. X-ray photoelectron spectroscopy (XPS) was used to analyze the Eu-ions' valence (Eu^{2+} or Eu^{3+}), equipped with a monochromatic source and a hemispherical electron energy analyzer.

3. Results and Discussion

XRD patterns of LSB-Eu phosphor synthesized at 850°C is presented in Fig. 1(a), which shows the characteristic peaks of the synthesized LSB-Eu powder. Diffraction peaks at 2θ values of 21.60°, 22.06°, 23.84°, 25.26°, 27.68°, 30.52°, 35.44°, 40.28°, 42.68°, 47.36°, 47.60°, 48.80°, 51.90° and

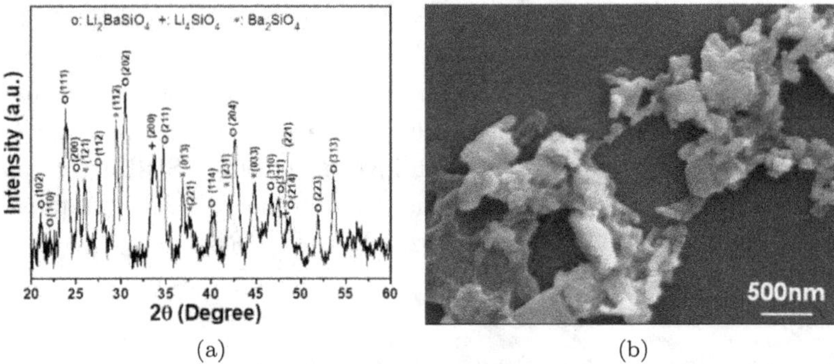

Fig. 1.   (Color online) (a) XRD patterns and (b) SEM images of Li_2BaSiO_4:0.003Eu^{2+} phosphor.

$53.62°$ were observed for the Li_2BaSiO_4 phase. The Li_2BaSiO_4 powder had a hexagonal crystal structure, which accords with the standard JCPDS No. 016-8709 card. Diffraction peaks for the Ba_2SiO_4 phase at 2θ values of $26.10°$, $29.54°$, $36.88°$, $37.60°$, $42.16°$ and $44.86°$ were also observed, which accords with the standard JCPDS No. 084-4279. Diffraction peaks for the Li_4SiO_4 phase at 2θ values of $33.84°$ and $48.72°$ were also observed, which accords with the standard JCPDS No. 037-1472. Kulshreshtha *et al.* had successfully synthesized $Li_2(Sr,Ca,Ba)SiO_4$:Eu^{2+} ternary composition library and they found several intermediate compounds (mixtures or solid solutions) were observed in a specific composition range.[2,3] When the synthesis temperature was 850°C, the LSB-Eu powder was shaped irregularly, as Fig. 1(b) shows, and the size increased to 0.25–1.25 μm. As we know, fine phosphor powders are difficult to obtain the improved PL characteristics. Because the sizes of micro-particulation are smaller than 2 μm, the surface defects will increase to cause degeneration of the emission intensity. However, even the average size of LSB-Eu particles is smaller than 1 μm, we show that it can be active as PL properties.

The PLE spectra in Fig. 2(a) show that as the RTRS changed, the measured spectrum revealed different results. The excitation spectra were measured at an emission wavelength of 501 nm. For LSB-Eu-800 phosphor, the emission intensities decreased with increasing excitation wavelength, and two emission peaks centered at ~ 242 nm and ~ 283 nm were observed. When the RTRS was lower than 800°C, the emission intensities increased with decreasing RTRS, and two emission peaks centered at ~ 283 nm and ~ 365 nm were observed. The results in Fig. 2(a) also shows an important

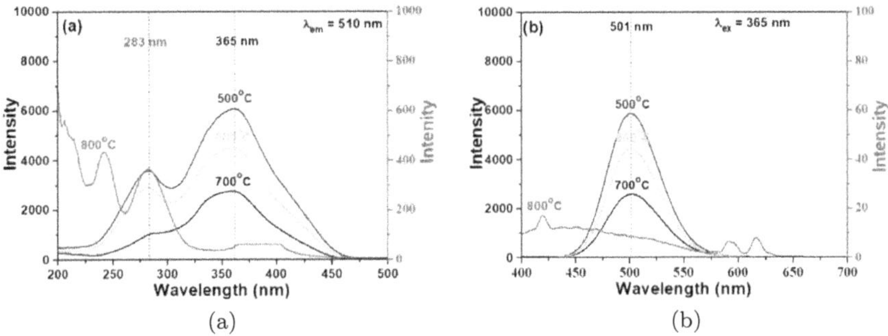

(a)　　　　　　　　(b)

Fig. 2. (Color online) (a) PLE spectra and (b) PL spectra of Li$_2$BaSiO$_4$:0.003Eu^{2+} phosphor as a function of removing temperature of reduction gas.

result that the maximum emission intensity of PLE spectrum for the LSB-Eu-800 phosphor is only one tenth of those with RTRS lower than 800°C.

These results prove that the RTRS is the most important factor that will affect the PLE and PL characteristics of LSB-Eu phosphors. From the results in Fig. 2(a) we found that the optimum excitation optical wavelength (λ_{ex}) for the LSB-Eu phosphor was around 365 nm for green light, and the LSB-Eu phosphor excited by other wavelengths had weaker PL intensities. The emission spectrum of LSB-Eu phosphor excited by UV light with the wavelengths of 283 nm or 365 nm is shown in Fig. 2(b). As the RTRS was 800°C, three weak emission peaks, which are located at 418 nm, 588 nm and 613 nm, were observed.

From the emission spectrum of LSB-Eu-800 phosphor, emission peaks at about 534 nm, 654 nm and 691 (and 706) nm corresponding to $^5D_0 \rightarrow {}^7F_0$, $^5D_0 \rightarrow {}^7F_3$, and $^5D_0 \rightarrow {}^7F_4$ transitions of Eu^{3+} ions were not found, and two weak emission peaks at about 588 nm and 613 nm corresponding to $^5D_0 \rightarrow {}^7F_1$ and $^5D_0 \rightarrow {}^7F_2$ transitions of Eu^{3+} ions were really found.[10,11] Because the 588 nm and 613 nm are the peaks of red light, which suggest that for LSB-Eu-800 phosphor, the reduced Eu^{2+} ions will be re-oxidized into Eu^{3+} ions. As the RTRS was lower than 800°C, one board-band green emission centered at ~ 501 nm was observed, which is assigned to the $4f^7 \rightarrow 4f^6 5d^1$ transition of Eu^{2+} ions.[12] It can be seen that the maximum emission intensity increased as the RTRS decreased from 700°C to 500°C and saturated at 500°C. These results prove that the RTRS influences the intensities of the emission spectra of LSB-Eu phosphors.

XPS is one kind of electron spectroscopy for chemical analysis (ESCA), and it can provide valuable quantitative and chemical state information

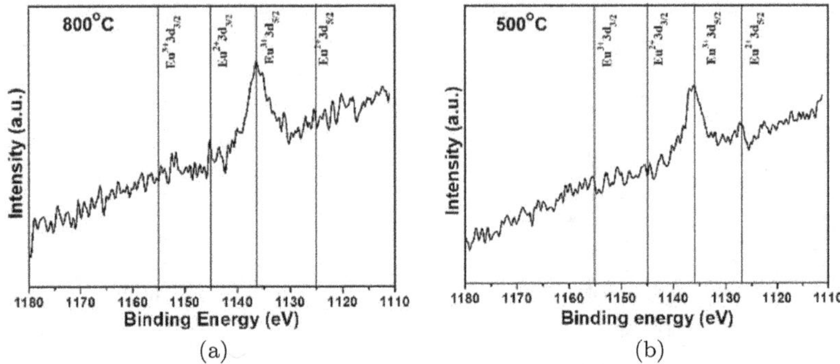

Fig. 3.   XPS spectra of Eu (3d) in Li_2BaSiO_4:$0.003Eu^{2+}$ phosphor at 300 K as a function of removing temperature of reduction gas (a) $800°C$ and (b) $500°C$.

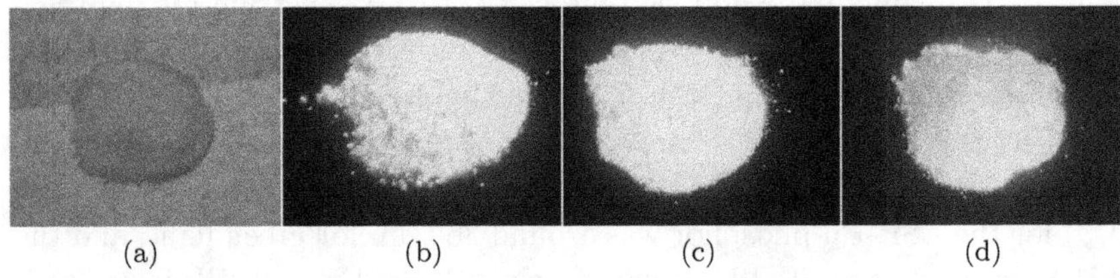

Fig. 4.   (Color online) Surface images of Li_2BaSiO_4:$0.003Eu^{2+}$ phosphors as a function of removing temperature of reduction gas (a) $800°C$, (b) $700°C$, (c) $600°C$ and (d) $500°C$.

from the surface of the material. For that, XPS was used to analyze the Eu-ions' valence of the LSB-Eu-800 and LSB-Eu-500 phosphors, and the results are shown in Fig. 3. From the results of XPS spectra we found for both LSB-Eu-800 and LSB-Eu-500 phosphors, the peaks of $Eu^{3+}3d_{3/2}$ and $Eu^{2+}3d_{3/2}$ do not exist, the peak position of $Eu^{3+}3d_{5/2}$ shifted to 1143 eV rather than at 1145 eV. $Eu^{2+}3d_{5/2}$ peak, which was located at 1127 eV, was not observed in LSB-Eu-800 phosphor but was really observed in LSB-Eu-500 phosphor.

Figure 4 presents photographs of LSB-Eu phosphors under the excitation of ultraviolet light as a function of RTRS. Even though the LSB-EU-800 phosphor has with weak emission peak, the photo image shows that the phosphor reveals a weak red color. When the RTRS is lower than $800°C$, the photo images of LSB-Eu phosphors reveal a bright green color. These photographs also prove that very bright green emission can easily be obtained by synthesizing LSB-Eu phosphors at $850°C$.

4. Conclusions

When the synthesis temperature of the LSB-Eu powder was 850°C, the Li_2BaSiO_4, Ba_2SiO_4 and Li_4SiO_4 phases were observed in the XRD pattern. The maximum emission intensity of PLE spectrum for the LSB-Eu-800 phosphor was only one tenth of those with RTRS lower than 800°C. Two weak emission peaks at 588 nm and 613 nm corresponding to $^5D_0 \rightarrow {}^7F_1$ and $^5D_0 \rightarrow {}^7F_2$ transitions of Eu^{3+} ions were really found in the LSB-Eu-800 phosphor. As the RTRS was in the ranges of 500–700°C, one boardband green emission centered at ~ 501 nm was observed, which is assigned to the $4f^7 \rightarrow 4f^65d^1$ transition of Eu^{2+} ions. The maximum emission intensity increased as the RTRS decreased from 700°C to 500°C. Compared to the XPS spectra, we found the $Eu^{2+}3d_{5/2}$ peak was not observed in LSB-Eu-800 phosphor but was really observed in LSB-Eu-500 phosphor. These results prove that the RTRS is an important to prepare the Li_2BaSiO_4:0.003Eu^{2+} green phosphor.

Acknowledgments

This work was supported by projects under Ministry of Science and Technology Grant Nos. MOST 108-2221-E-390-005 and MOST 108-2622-E-390-002-CC3.

References

1. X. Y. Li, D. Y. Zhang, Y. Chen and C. K. Chang, *Ceram. Inter.* **43**, 1677 (2017).
2. C. Kulshreshtha, N. Shin and K. S. Sohn, *Electrochem. Solid-State Lett.* **12**, J55 (2009).
3. C. Kulshreshtha, A. K. Sharma and K. S. Sohn, *J. Electrochem. Soc.* **156**, J52 (2009).
4. X. M. Feng, X. S. Wang, J. Li, Y. Zhang and X. Yao, *J. Mater. Sci. Mater. Electron.* **28**, 7177 (2017).
5. G. Blasse and B. C. Grabmaier, *Luminescence Material* (Springer, Berlin, 1994).
6. J. Liu, C. C. Wu, C. F. Yang and L. S. Liou, *Nano* **14**, 1950110 (2019).
7. J. Liu, K. N. Chen, W. C. Tzou, M. P. Houng, Y. I. Ho and C. F. Yang, *Sensors Mater.* **29**, 473 (2017).
8. Z. Wei, Y. L. Wang, X. B. Zhu, J. N. Guan, W. X. Mao and J. J. Song, *Chem. Phys. Lett.* **648**, 8 (2016).

9. H. He, R. L. Fu, X. F. Song, R. Li, Z. W. Pan, X. R. Zhao, Z. H. Deng and Y. G. Cao, *J. Electrochem. Soc.* **157**, J69 (2010).
10. C. Y. Lin, S. H. Yang, J. L. Lin and C. F. Yang, *Appl. Sci.* **7**, 30 (2017).
11. K. N. Chen, C. M. Hsu, J. Liu, Y. T. Chiu and C. F. Yang, *Appl. Sci.* **6**, 22 (2016).
12. M. P. Saradhi and U. V. Varadaraju, *Chem. Mater.* **18**, 5267 (2006).

CNT/epoxy interleaves for strengthening performance of filament wound composite structures*

Soo-Jeong Park[†] and Yun-Hae Kim[‡]

*Department of Ocean Advanced Materials Convergence Engineering,
Korea Maritime and Ocean University, Busan, 49112, Republic of Korea*
[†] *blue9069@naver.com*
[‡] *yunheak@kmou.ac.kr*

The failure mechanism of composites dominates the matrix, fiber and interface, and in general, the matrix corresponds to the definitive cause of damage. A filament–wound composite structure involves a notable bridging effect owing to the matrix between the layers, and particle additives are generally adopted to strengthen the matrix. However, particle additives exhibit a low performance when applied to structures, owing to the dispersibility and particle agglomeration. In this study, the strengthening performance of carbon nanotube (CNT)/epoxy interleaves was experimentally verified to facilitate their implementation in the structural design of a filament–wound cylinder structure. The burst pressure, compression, bending and interfacial bonding strength of the cylinder improved by approximately 20%, 161%, 16% and 36%, respectively, and the positioning of CNT/epoxy interleaves was a more notable influencing factor compared to the proportion of CNTs in the entire winding layer. The number of macro voids decreased inside the epoxy modified CNT. The findings demonstrated that the incorporation of CNTs through CNT/epoxy interleaves could facilitate the matrix strengthening and enhance the interfacial bonding.

Keywords: CNT/epoxy; interleaves; filament; interfacial bonding.

1. Introduction

The failure mechanisms of filament–wound composite cylinders, which have an axisymmetric structure, dominate the matrix, fiber reinforcement and

---

[‡]Corresponding author.
*To cite this article, please refer to its earlier version published in the *International Journal of Modern Physics B*, Volume 35, 2140001 (2021), DOI: 10.1142/S0217979221400014.

matrix–fiber interfacial interaction.[1] Moreover, crack propagation, delamination and final fiber breakage may occur owing to the matrix damage.[1,2] Many researchers have attempted to address this problem, for instance, by changing the stacking sequence to control the crack growth by changing the initial crack positioning and to delay the crack formation by adding nanofillers.[2] Carbon nanotubes (CNTs), owing to their excellent fracture toughness and load transferability, can facilitate crack bridging and increase the crack initiation energy by upto 75%.[1,3–5] However, CNTs are industrially limited owing to their dispersibility and particle agglomeration and exhibit a low production efficiency and economics disadvantages for structures with certain forms. Considering these aspects, in this study, the use of CNT/epoxy interleaves is considered to develop a structurally efficient fiber–matrix interface. It is expected that through interleaf winding, the brittleness of carbon fiber reinforced plastic (CFRP) can be minimized, and its resistance to crack propagation can be enhanced.[1] Moreover, the robustness of the cylinder structure against mechanical loads can likely be enhanced by increasing the fiber–matrix interfacial bonding strength locally through the CNT implementation.

2. Experimental Works

The adopted CFRP consisted of carbon filaments (T700S 12K from TORAYCA®), and bisphenol-A-based epoxy resin (KFR-120V KUKDO Chemical Co.; KRH-141 hardener with a 100:30 ratio), fiber reinforcement, and matrix. Nano additives were modified to epoxy using NanoLab, Inc.'s multiwall CNTs. These materials have a hollow structure with a purity rating of $> 95\%$, and an average length, outer diameter and specific surface area of 5–20 μm, 15 ± 5 nm and 200–400 m^2/g, respectively. CNT–strengthened CFRP specimens having two shapes were produced, as summarized in Table 1. As shown in Fig. 1, the first type had a cylinder structure, and the second type had a flattened structure realized by conducting vacuum-assisted forming after cutting the filament–wound cylinder in the longitudinal direction before curing. The fiber weight fraction of the specimens was measured according to the ASTM D3171. To analyze the strengthening effect of the CNT/epoxy interleaves, tests to determine the tensile, flexural, interlaminar shear strength (ILSS), and parallel-plate loading behavior (ASTM D3039, D790, D2344, D2412, respectively) based on the burst pressure were performed.

Table 1. Summary of all the test specimens and their design; E: neat epoxy, CE: epoxy modified with CNTs.

Specimen label	CNTs content (wt.%)	Winding angle (°)	Layer sequence*	Winding diameter (mm) Cylinder (x = C)	Flat (x = F)	Fiber weight fraction (wt.%)
x-E3	N/A	75	$[E]_3$	50	76.2	81.8 ($\pm$0.2)
x-E2CE1	0.1/1ply	75	$[E_2/CE_1]_1$	50	76.2	80.9 ($\pm$0.2)
x-E1CE1E1	0.1/1ply	75	$[E/CE/E]_1$	50	76.2	80.4 ($\pm$0.3)
x-CE1E2	0.1/1ply	75	$[CE_1/E_2]_1$	50	76.2	82.8 ($\pm$0.1)
x-CE3	0.1/1ply	75	$[CE]_3$	50	76.2	81.3 ($\pm$0.6)

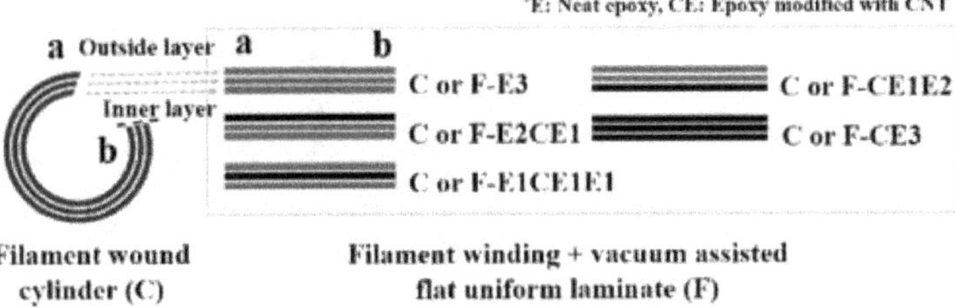

Fig. 1. (Color online) Two typed structures of the CNT/epoxy interleaved CFRP composites.

3. Results and Discussion

Figure 2(a) shows the results for the burst pressure and elastic modulus. When a load is applied from the inside to the outside, CNT/epoxy interleaves is approximately 20% more than the burst pressure of C-E1CE1E1 compared to C-E3. Thus, a strengthening effect is noted for C-CE3, which is about 10% higher than that of C-CE3. The elastic modulus is majorly influenced by the proportion of CNT/epoxy interleaves in the structure. However, the positioning of the CNT/epoxy interleaves more notably influences the strength. The presence of the CNT/epoxy interleaved increased the interfacial bonding strength against the shear force generated in the interface layer. In other words, these structures can suppress the initial crack formation and effectively delay the failure. The tensile test results demonstrate that the burst pressure is more dominant in the fiber, and misalignment occurs because of the layer-separated filament winding of the CNT/epoxy interleaves.

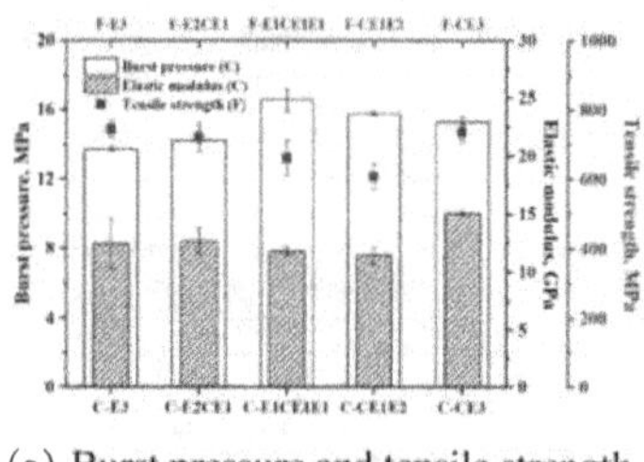

(a) Burst pressure and tensile strength

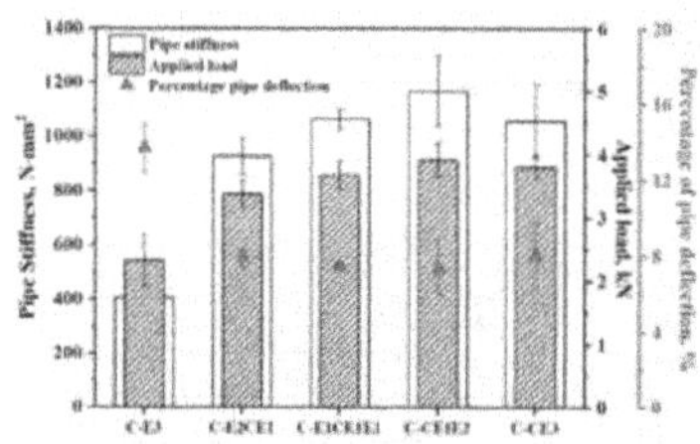

(c) Parallel-plate loading behavior

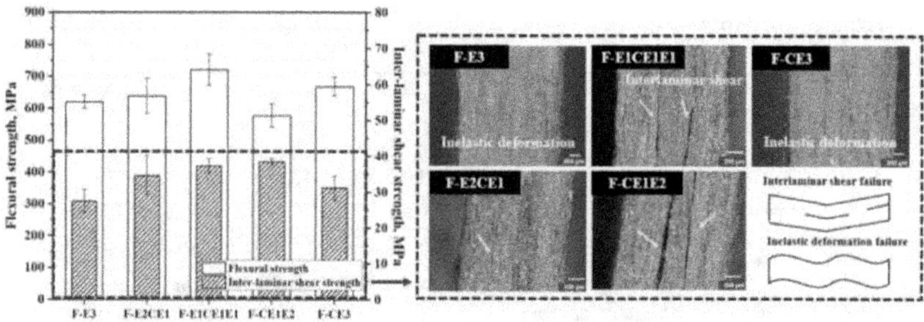

(b) Flexural and ILSS with failure modes by optical microscope

Fig. 2.   (Color online) Strengthening performance on CNT/Epoxy interleaves.

Figure 2(b) illustrates the effect of the matrix on the failure mechanism and the failure modes in the short beam test. A larger number of CNT/epoxy leaves located in the interface layer closer to the outside corresponds to a higher flexural strength. The strength of F-CE3 and F-E2CE1 is approximately 7% and 16% higher than that of F-E3, respectively. This finding indicates that even a small amount of CNTs leads to a notable strengthening effect, thereby enhancing the interfacial bonding strength. This effect can be enhanced by interleaving CNTs/epoxy close to the point at which the external load acts. Preventing cracking on the surface is more effective than suppressing the crack propagation. The ILSS of F-CE1E2, F-E1CE1E1, F-E2CE1 and F-CE3 was approximately 40%, 36%, 26% and 13% more optimal than that of F-E3, respectively. Figure 2(c) shows the parallel-plate loading behavior obtained by evaluating the cylinder's strength and flexibility against complex loads. In general, the cylinder changes to an ellipse under parallel-plate loading, and stress concentration occurs in the transverse direction and contact area exposed to the direct load. When the structural deformation reaches the critical point of a stable equilibrium region, a sudden collapse occurs. The CNT/epoxy interleaves

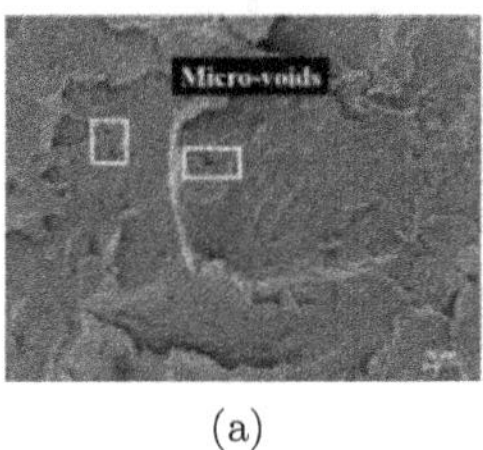

(a)

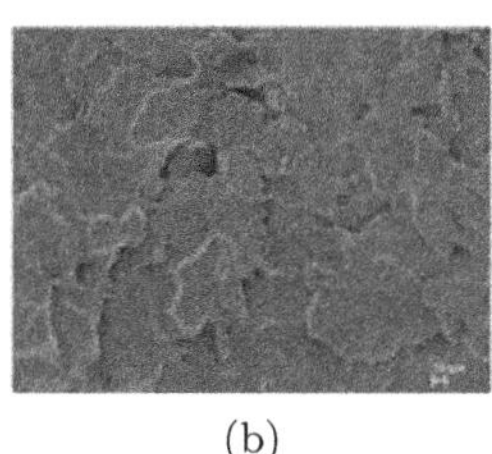

(b)

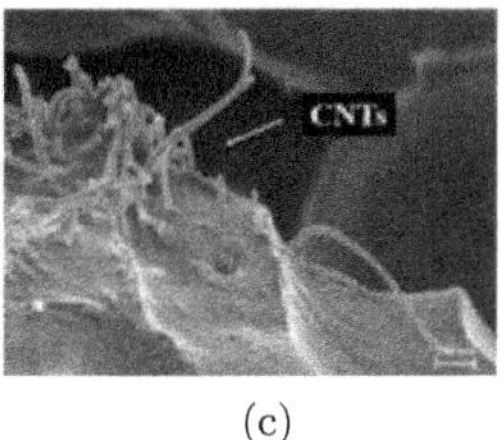

(c)

Fig. 3. SEM observation of (a) neat epoxy, (b) epoxy modified with CNTs and (c) CNTs agglomeration in epoxy.

dramatically enhance the stiffness by more than two times compared to that of C-E3 in all the considered cases. The stiffness of C-E1CE1E1 is approximately 161% higher than that of C-E3, and the deflection is approximately 54%, half the value of C-E3. Figures 3(a) and 3(b) show the scanning electron microscope images of the epoxy without and with CNTs. The fine pores inside the epoxy can be eliminated by the addition of CNTs, which indicates that CNTs lead to a stable interface of the fiber–matrix. In addition, as shown in Fig. 3(c), CNTs exhibit a partial aggregation inside the epoxy. This finding highlights the necessity of incorporating CNT/epoxy interleaves.

4. Conclusion

In a filament–wound CFRP cylinder structure, the incorporation of CNTs helped improve the elastic modulus. However, in terms of strengthening, the positioning of the CNT/epoxy interleaves was more important. Even a small amount of effectively designed CNT/epoxy interleaves can lead to sufficient strengthening. Furthermore, the CNT/epoxy interleaves could prevent crack formation on the surface, which is more effective than suppressing the crack propagation. The stiffness improved by at least 128% and at most 187% in terms of the parallel-plate loading behavior. Nevertheless, the layer-separated filament winding process may lead to a misalignment between the layers, and thus, additional research on the process of dominant fiber breakage must be performed.

Acknowledgment

This work was supported by the Technology Innovation Program (No. 20005403), funded by the Ministry of Trade, Industry & Energy (MOTIE, Korea).

References

1. T. Üstün, H. Ulus, S. E. Karabulut, V. Eskizeybek, Ö. S. Şahin, A. Avci and O. Demir, *Compos. B Eng.* **96**, 1 (2016).
2. J. H. S. Almeida Jr, M. L. Ribeiro, V. Tita and S. C. Amico, *Compos. Struct.* **160**, 204 (2017).
3. P. Lv, Y. Y. Feng, P. Zhang, H. M. Chen, N. Zhao and W. Feng, *Carbon* **49**, 4665 (2011).
4. M. T. Kim, K. Y. Rhee, J. H. Lee, D. Hui and A. K. Lau, *Compos. B Eng.* **42**, 1257 (2011).
5. Y. T. Jung, H. D. Roh, I. Y. Lee and Y. B. Park, *Funct. Compos. Struct.* **2**, 015006 (2020).

Carbon nanotubes produced from rubber fruit shell activated carbon*

Suhdi[†] and Sheng-Chang Wang[‡]

*Department of Mechanical Engineering, Southern Taiwan University
of Science and Technology, Tainan City 71005, Taiwan*
[†] *da61y203@stust.edu.tw*
[‡] *scwang@stust.edu.tw*

This study used rubber fruit shells (RFS) as raw material for making carbon nanotubes (CNTs). The CNTs were carried out by hydrothermal process at low-temperature after carbonization and chemical activation was done. This experiment succeeded in obtaining a bundle of MWCNT (Multi-Wall Nanotube) from the raw material of RFS. The results of characterization using SEM and TEM showed that the resulting CNTs were not homogeneous in diameter, ranging from 13 to 455 nm, with an average diameter of about 179 nm. XRD was used to identify crystallographic structure; it has two peaks 2θ at around 26.0 and 44.0, index to 002 and 101 reflections hexagonal graphite diffraction of the MWCNTs. This study can provide an alternative inexpensive raw material and a simple method to obtain carbon nanotubes.

Keywords: Rubber fruit shell; multi-wall nanotube; hydrothermal process; crystallographic structure.

1. Introduction

Currently, carbon nanotubes (CNTs) can be used in many applications for engineering such as structural reinforcement,[1,2] coatings,[3] electromagnetics,[4,5] sensors and actuators applications,[6] hydrogen storage,[7] and also for biotechnological and biomedical applications.[8,9] Three common techniques are arc discharge technique,[10,11] laser ablation technique.[10,12] and chemical vapor deposition technique.[13] Several materials for the CNTs synthesis have

---

[‡]Corresponding author.
*To cite this article, please refer to its earlier version published in the *International Journal of Modern Physics B*, Volume 35, 2140003 (2021), DOI: 10.1142/S0217979221400038.

recently been used. One of the critical materials for synthesizing carbon materials is biomass; biomass is a qualified carbonaceous because it is available in high quality and enormous quantities and is an environmentally friendly renewable resource.[14] Several data have shown that precious CNTs can be obtained from biomass feeds, such as coconut shells using pyrolysis methods,[15] bamboo charcoal by pyrolysis methods in ethanol vapor,[16] pine nutshell at low-temperature via microwave-assisted CVD using Ni catalyst,[17] wood sawdust in a tabular reactor with Fe/Mo/Mg O as a catalyst.[18] However, high reaction temperatures and metal catalysts are always required for most of these processes. In this research, rubber fruit shells (RFS) as new natural materials have been used to get CNTs. The RFS as a precursor has carbonized using the pyrolysis process continued with the chemical activation using KOH solution, and then to obtain CNTs, the rubber fruit Shell activated carbon (RFSAC) are carried out hydrothermal process at low-temperature and without adding a metal catalyst.

2. Material and Experimental

The RFS from Bangka Island of Indonesia was used to prepare the activated carbon as a substrate to obtain CNTs. The chemical composition of RFS, such as cellulose, lignin and pentosane, is 48.68%, 33.54% and 16.81%, respectively.[19] The RFS was washed with distilled water and soaked for 4 h using 10% H_2SO_4 solution, and then dried for 24 h using an oven at 60°C. The dried materials were carbonized in a furnace at 450°C with 20 mL/min nitrogen gas and held for 1 h. Finally, the carbonized charcoal from RFS was crushed manually using a pestle and mortar and then sieved to get 200 mesh/74 μm of charcoal powder.

The 3 g of RFS charcoal powder was mixed with KOH solution in a 100 mL beaker with the composition weight ratio of 1:4. The solution was then mixed at 1000 rpm for 1.5 h until it became homogenous slurry and then dried for 12 h in an oven at 110°C. Subsequently, the dry slurry was transferred in a ceramic crucible and pyrolysis process is done at 800°C in a furnace for 2 h.[20] The activated carbon was cooled in the furnace. It was neutralized at pH 7 with 1 M HCl and then washed with distilled water to eliminate the ionic contaminants.[21] The RFS activated carbon was subsequently immersed in 40 mL of distilled water and stirred for 3 h at 60°C, 300 rpm, then placed at room-temperature for one day. The final process was drying in the dry oven at 90°C until it completely dried which took few days.

The crystallographic structure was examined using XRD (Bruker-D2 Phaser) analysis. The morphology and microstructure on the surface were determined by Scanning Electron Microscope-Energy Dispersive Spectroscopy (JEOL JSM-6701F), and the Transmission Electron Microscopy (JEOL-JEM-1230).

3. Results and Discussion

Figure 1(a) shows the SEM images of the CNT grown on activated carbon surfaces. The structure and dimensions of CNTs are more clearly shown in TEM images in Fig. 1(b). It was a multi-walled carbon nanotube (MWCNT),[22] where the length and diameter size of CNTs are inhomogeneous and it is distinctly visible.

The histogram diagram and the Gaussian curve of the distribution of CNTs (Fig. 2) showed that the curve is asymmetric. It means that the distribution of the CNT on RFSAC is nonhomogeneous. The smallest size is 13 nm, and the largest is 455 nm and most commonly formed CNTs are between 150 and 200 nm in diameter and the average CNT diameter size distribution is around 179 nm. The primary peaks related to the MWCNTs have been clearly defined by XRD [Fig. 2(b)]. It has two peaks 2θ at around 26.0 and 44.0, index to 002 and 101 reflections hexagonal graphite diffraction planes, in good accordance with JCPDS card file, no. 41-1487[23] as shown in Fig. 2(b).

In several papers, 3d-valence transition metals Fe, Ni and Co, precious metals (e.g., Au) and poor metals (e.g., In, Pb) have been used as a catalyst to produce carbon nanotubes.[24] The morphology and quality of CNTs rely

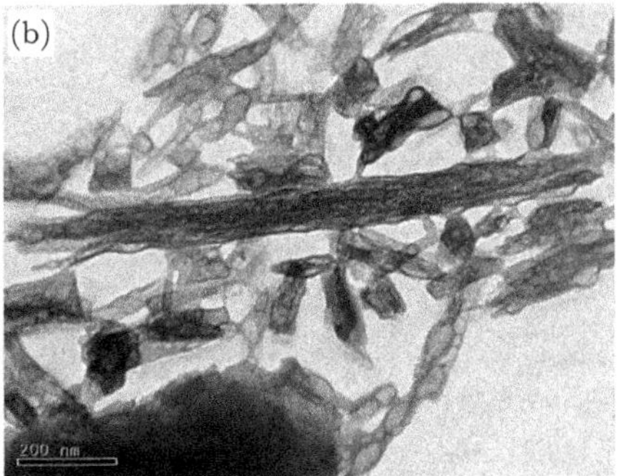

Fig. 1. (a) SEM micrographs of CNTs (magnification 5000x) and (b) TEM micrographs of CNTs.

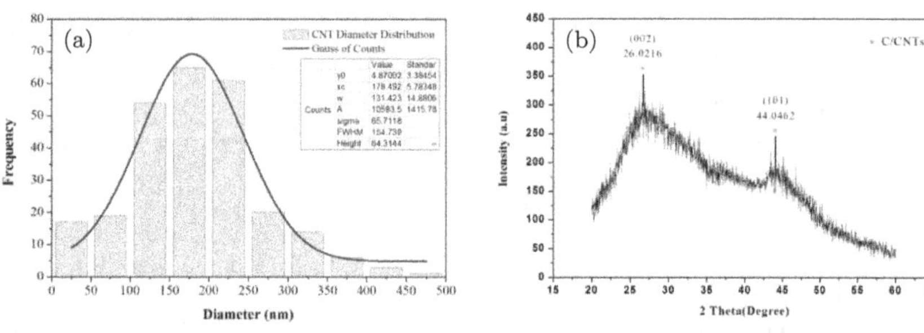

Fig. 2.   (Color online) (a) The histogram curve of CNTs diameter distribution and (b) the XRD pattern of CNTs.

heavily on the properties and form of composition of the catalyst.[25] In this work, the SEM and TEM result was apparent that although do not use a catalyst, the CNTs have resulted. One thing that might support the formation of CNTs in this study is the chemical composition contained in RFS. The RFS has 48.64% celluloses,[19] one of the lignocellulosic materials within plants can be easily altered into nano-sized fiber owing to elemental fibril of the cellulose linked and hydrogen bond.[26] Furthermore, according to Goodell *et al.*,[27] nanochannels developed at relatively low carbonization temperatures from the preferential ablation of cellulose microfibrils in the cell wall structure are proposed to assist in the catalysis of the CNTs. This result can be explained because RFS is a lignocellulosic material so that it can be easily converted into a CNT.

4. Conclusion

It has been proven that the CNTs can be obtained from RFSAC by carrying out hydrothermal process at low-temperature and without adding a metal catalyst. The primary peaks related to the MWCNTs have been clearly defined by XRD, it has two peaks 2θ at around 26.0 and 44.0, index to 002 and 101 reflections hexagonal graphite diffraction planes. SEM and TEM characterization of CNTs has inhomogeneous diameter size bundles; the diameter size ranges between 13 and 455 nm, with average diameter size 179 nm. Although the diameter size is inhomogeneous, this result proves that RFSAC can be recommended for CNTs material, and it can be used for any engineering applications.

References

1. T. Manzur, N. Yazdani and M. A. B. Emon, *Nanomater.* **2016**, ID 1421959 (2016).
2. K. M. Liew, M. F. Kai and L. W. Zhang, *Compos. A Appl. Sci. Manuf.* **91**, 301 (2016).
3. L. Calabrese, A. Khaskoussi and E. Proverbio, *Fibers* **8**, 57 (2020).
4. M. Zakariah and P. Puspitasari, *J. Mech. Eng. Sci. Technol.* **1**, 38 (2017).
5. L. Kong *et al.*, *J. Mater. Chem. C* **5**, 7479 (2017).
6. P. Chandrasekhar, in *Conducting Polymers, Fundamentals and Applications 53–60* (Springer International Publishing, 2018).
7. E. Anikina *et al.*, *ACS Appl. Nano Mater.* **2**, 3021 (2019).
8. T. Saliev, *C* **5**, 29 (2019).
9. K. Kuche *et al.*, *Nanoscale* **10**, 8911–8937 (2018).
10. R. Das *et al.*, *Nanoscale Res. Lett.* **11**, 510 (2016).
11. A. H. Sari *et al.*, *Int. Nano Lett.* **8**, 19 (2018).
12. F. V. Ferreira *et al.*, *Carbon Nano-Objects* (Elsevier Inc., 2018).
13. N. Duc Vu Quyen *et al.*, *J. Chem.* **2019**, ID 4260153 (2019).
14. M. A. Perea-Moreno *et al.*, *Sustain.* **11**, 863 (2019).
15. A. Melati and E. Hidayati, *J. Phys. Conf. Ser.* **694**, 012073 (2016).
16. N. Arnaiz *et al.*, *J. Anal. Appl. Pyrolysis* **130**, 249 (2018).
17. J. Zhang *et al.*, *Diam. Relat. Mater.* **91**, 98 (2019).
18. M. G. S. Bernd *et al.*, *J. Mater. Res. Technol.* **6**, 171 (2017).
19. M. Meilianti, *J. Distilasi* **2**, 1 (2018).
20. G. Asimakopoulos *et al.*, *Appl. Sci.* **10**, 1 (2020).
21. S. Li, *Int. J. Electrochem. Sci.* **13**, 1728 (2018).
22. S. Zhang *et al.*, *RSC Adv.* **4**, 54244 (2014).
23. N. Yahya *et al.*, *Nanoscale Res. Lett.* **14** (2019).
24. M. H. Rümmeli *et al.*, *Nanoscale Res. Lett.* **6**, 1 (2011).
25. S. S. Madani *et al.*, *Int. J. Nano Dimens.* **7**, 240 (2016).
26. E. Azwar *et al.*, *Int. J. Hydrogen Energy* **43**, 20811 (2018).
27. B. Goodell *et al.*, *J. Nanosci. Nanotechnol.* **8**, 2472 (2008).

Fabrication of strong macrofibers from plant fiber bundles*

Antonio Norio Nakagaito[†] and Hitoshi Takagi[‡]

*Graduate School of Technology, Industrial and Social Sciences,
Tokushima University, 1-1 Minamijosanjima-cho,
Tokushima-shi, Tokushima-ken 770-8506, Japan*
[†] *nakagaito@tokushima-u.ac.jp*
[‡] *takagi@tokushima-u.ac.jp*

Yusuke Katsumoto

*Graduate School of Advanced Technology and Science,
Tokushima University, 1-1 Minamijosanjima-cho,
Tokushima-shi, Tokushima-ken 770-8506, Japan*
yksoccer8mf@gmail.com

The production of long cellulose macrofibers starting from nanofibers is still complex and expensive. This study proposes a method of manufacturing long macrofibers from plant fiber bundles by chemically extracting non-cellulosic substances but preserving the original shape of the fibers. Once the cellulosic fiber bundles are dried, the original cellulose nanofibers are bridged to neighboring nanofibers by the formation of hydrogen bonds, giving shape stability and enhanced mechanical properties. By the process, the tensile strength was increased by about 50% and the modulus doubled from the original plant fiber bundles.

Keywords: Cellulose; nanofiber; macrofiber; strength.

1. Introduction

The manufacture of continuous fibers aiming at a variety of uses such as in textiles, filtration elements and composite reinforcements, is in great demand. Continuous fibers are particularly attractive due to the possibility

---

[†] Corresponding author.
*To cite this article, please refer to its earlier version published in the *International Journal of Modern Physics B*, Volume 35, 2140005 (2021), DOI: 10.1142/S0217979221400051.

of controlling fiber orientation when fabricating composites, and many high-performance composites are based on strong fibers such as carbon, aramid or glass. However, besides having high mechanical properties, materials must also attend environmental requirements both during manufacture and at end-of-life disposal. To attend these new demands, natural fiber bundles from many plants have been converted into yarns and used as reinforcements in composites. However, the mechanical properties are much lower than those of synthetic fibers and are dependent on variations in climate and cultivation conditions to what fiber-producing crops were subjected.

The latest trend in bio-based materials for composite manufacture has been the exploration of cellulose nanofibers extracted from plant fibers. They are made up of long cellulose molecular chains forming a semicrystalline arrangement that confers tensile modulus, strength and density on par with synthetic aramid.[1] Even though having remarkable mechanical properties (the tensile modulus was measured as 138 GPa[2] and the strength is estimated to be 1.6–3 GPa[3]), controlling the proper alignment of nanofibers has been one of the major challenges. Many successful approaches are being proposed such as fiber spinning and hydrodynamic alignment as in Ref. 4, but they are still complex and costly. One interesting approach proposed by Wang *et al.*[5] obtained high-strength/modulus macrofibers by twisting and drying wet bacterial cellulose strips. The constituting nanofibers would be fixed in shape exclusively by hydrogen bonds. In this study, instead of reforming fibers from nanofibers, Manila hemp fiber bundles had lignin and hemicelluloses chemically removed and subsequently rinsed with water while keeping the integrity of the original fiber bundle shape. Wet fiber bundles were twisted and dried under tension to form long fiber bundles comprised of virtually pure cellulose. The tensile strength was increased by more than 50% and tensile modulus more than doubled relative to the original fiber bundles.

2. Materials and Methods

2.1. *Materials*

Long Manila hemp fiber bundles were used as the starting material. Sodium chlorite and acetic acid were purchased from Kanto Chemical Co., and potassium hydroxide from FUJIFILM Wako Pure Chemical Corporation.

2.2. *Lignin removal from fiber bundles*

A mixture of 360 mL of distilled water, 2.4 g of sodium chlorite and 0.48 mL of acetic acid was prepared and kept at 75°C, in which about 6 g of Manila hemp fiber bundles were immersed. The same amounts of sodium chlorite and acetic acid were added to the mixture every 1 h of treatment, totalizing 3 h of treatment to completely remove lignin. Fibers were supported by a mesh and carefully rinsed with running water until pH became neutral.

2.3. *Yarn spinning, hemicelluloses removal and drying*

Treated fiber bundles still in wet state were cut in lengths of about 100 mm, twisted and fixed in their edges to keep them straight. Extended fibers were then immersed in a 6 wt.% aqueous solution of potassium hydroxide at room-temperature for 12 h and at 80°C for 2 h to remove hemicelluloses. Yarns were rinsed again in running water and dried in a convection oven at 105°C.

2.4. *Tensile test and optical microscopy observations*

Tensile tests were performed using a universal materials testing machine Instron Model 5567 (Instron Corp., USA) at a crosshead speed of 1 mm/min and a gauge length of 25 mm. Fiber bundles and fabricated yarns were observed by a field emission scanning electron microscope Hitachi S-4700 (Hitachi High-Tech Corp., Japan).

3. Results and Discussion

The results of tensile test are shown in Fig. 1. The chemical treatment of fiber bundles produced a significant gain in tensile properties. The untreated fiber bundle had about 380 MPa and 18 GPa of strength and modulus, respectively, by which increased to 680 MPa and 49 GPa when chemically treated. These values compare well with fibers spun from cellulose nanofiber (383.3 MPa, 23.9 GPa),[6] or from bacterial cellulose nanofiber (357.5 MPa, 22.9 GPa)[7] suspensions. However, when twisting was applied, the mechanical properties decreased monotonically with the number of twisting turns. The enhanced tensile properties achieved by chemical treatment is explained by the removal of lignin and hemicelluloses, leaving only stronger cellulose nanofibers bridged by hydrogen bonds formed during drying. Besides, the lumina present in the untreated fibers [Fig. 2(a)] were

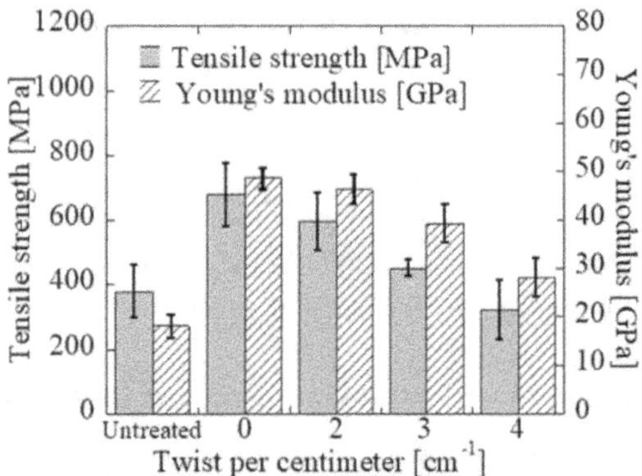

Fig. 1.   Tensile properties of untreated Manila hemp fiber bundle and derived yarns obtained at varying twisting turns.

Fig. 2.   Cross section of untreated Manila hemp fiber bundles showing open lumina (a) and of a chemically treated yarn with collapsed lumina (b).

collapsed after treatment, increasing the density of the yarns. On the other hand, twisting led to deviation in the alignment of individual fibers making an angle with the length of the yarns, as depicted in Fig. 3. As most of the nanofibers are aligned longitudinally to the fibers, their alignment accompanies that of fibers and therefore results in tensile strength decrease of the obtained macrofibers. By removing non-cellulosic substances without twisting the yarns, most of the cellulose nanofibers were kept aligned along the length of the macrofibers due to the previous natural arrangement in the thickest layer (called S2 layer) of the cell wall of each fiber.

Considering the typical strength of Manila hemp fiber bundles, which is generally higher than the value reported here, the gain in strength by the present process was not so significant. Yet, this process might be able to

Fig. 3. Untreated Manila fiber bundle showing fibers aligned longitudinally (a) and chemically treated and twisted yarn showing fibers making an angle with the yarn length (b).

keep the mechanical properties consistent regardless of the property variations inherent to plant fibers.

4. Conclusion

The proposed process allows the fabrication of macrofibers with most of the cellulose nanofibers axially aligned, without the need of nanofibrillation and subsequent realignment required by other processes. The method is simple and the tensile properties obtained were superior to those of the original plant fiber bundles.

References

1. J. E. Gordon, *The New Science of Strong Materials* (Princeton University Press, Princeton, 1976).
2. T. Nishino, K. Takano and K. Nakamae, *J. Polym. Sci. Pol. Phys.* **33**, 1647 (1995).
3. T. Saito *et al.*, *Biomacromolecules* **14**, 248 (2013).
4. C. Clemons, *J. Renew. Mater.* **4**, 327 (2016).
5. S. Wang *et al.*, *Adv. Mater.* **29**, 201702498_1 (2017).
6. A. Kafy *et al.*, *Sci. Rep. – UK* **7**, 17683_1 (2017).
7. J. Yao *et al.*, *ACS Appl. Mater. Interfaces* **9**, 20330 (2017).

A study on the removal of paint and oxide layer on the steel surface by laser beam scanning method*

Ji-Eon Kim

*Ocean Science and Technology School,
Korea Maritime and Ocean University, 727 Taejong-ro,
Yeongdo-gu, Busan 49112, South Korea
wldjs010@naver.com*

Jong-Do Kim[†]

*Division of Marine Engineering,
Korea Maritime and Ocean University, 727 Taejong-ro,
Yeongdo-Gu, Busan 49112, South Korea
jdkim@kmou.ac.kr*

In this study, a laser cleaning method was developed in order to replace the existing manual work-based technologies. The experimental investigation analyzed the cleaning properties of the paint and the oxide layer on the steel surface according to the laser beam scanning method. Experiments showed that in the cleaning process, the line beam patterns with a moving stage showed differences in heat input depending on the area, and the square area beam patterns obtained with control of a Galvano scanner showed uniform cleaning performance. The results of this study demonstrated that through the prosed laser cleaning technique, oxide layers as well as paint on steel surfaces can be removed with excellent precision, which is expected to be useful in the development of eco-friendly surface cleaning techniques.

Keywords: Laser cleaning process; eco-friendly technology; application for shipbuilding.

1. Introduction

Marine powerhouse such as Korea, Japan, China and Europe are pushing ahead for a smart yard aiming for building optimal logistics and

---

[†]Corresponding author.
*To cite this article, please refer to its earlier version published in the *International Journal of Modern Physics B*, Volume 35, 2140006 (2021), DOI: 10.1142/S0217979221400063.

shipbuilding system as well as smart shipyards with world-class productivity. The process of shipbuilding includes fabrication of parts and piping, assembly of flat panel blocks and curve blocks, outfitting, painting and erection. This study aims to develop eco-friendly painting pretreatment and surface cleaning techniques. By replacing the existing manual work-based technologies that often entail problems in the safety of workers and environment with a laser cleaning method, productivity is enhanced through decreased lead-time, working environment is improved and discharging of environmental pollutants can be significantly reduced.[1-5] The experiment was conducted with laser beam scanning method as the main process parameter considering the on-site working environment, and characteristics on cleaning of the anti-fouling paint and oxide layer on the steel surface were analyzed.

2. Experimental Material and Method

As the experimental materials, a steel plate coated with anti-fouling paint and an oxidized steel plate were used, and the thickness of the paint and oxide layer was 170 and 110 μm, respectively.

Table 1 shows experimental conditions using portable type laser cleaning equipment. The main process parameters of this study are the beam patterns according to the laser beam scanning method. Figure 1 shows a schematic diagram of the experimental method with the Line and Square area beam patterns. In each experiment, the scan area of the specimen surface was 50 mm × 50 mm. When a line beam pattern is used in the field, the workpiece must be fixed and the human operator was manually moved with the position of the laser head. However, in this study, for quantitative analysis of the experimental results, the stage on which the specimen was placed moved by 50 mm in y-axis direction with the laser head fixed.

Table 1.   Experimental conditions.

Beam pattern	Line	Square area
Average power	100 W	
Energy density	13.6 J/cm^2	
Pulse overlap rate	50%	
Line overlap rate	50%	
Stage speed	7 mm/s	—

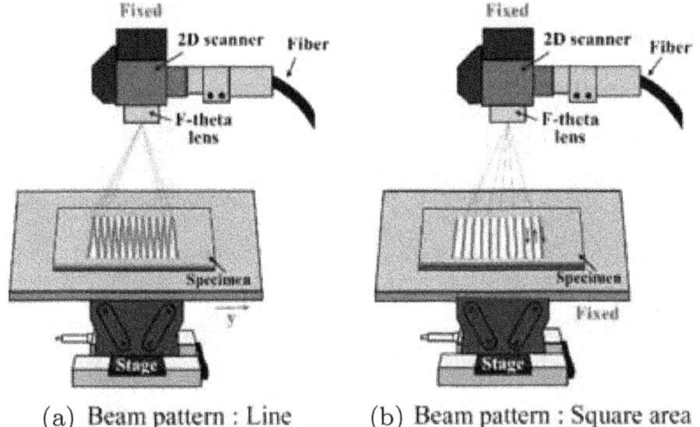

(a) Beam pattern : Line   (b) Beam pattern : Square area

Fig. 1. (Color online) Schematic of laser cleaning experimental method by the laser beam pattern.

In case of using Square area beam patterns, laser beam scans the area input in the system by movement of the Galvano scanner inside the laser head, while work piece and laser head are fixed.

3. Results and Discussion

3.1. *Comparison of paint removal characteristics*

Figure 2 shows the images of the laser cleaned surface according to the beam pattern type and the number of scans. When the laser beam is applied on the specimen surface, the temperature rises and the anti-fouling paint on the steel surface evaporates and decomposes. Therefore, the paint layer was removed in five laser scans for each laser beam pattern.

Comparing the surface photographs of the cleaned area according to the beam pattern, in case of Line beam patterns, the paint is removed first in the upper and lower edge areas. On the other hand, the paint was removed uniformly over the entire area when the Square area beam pattern was used. These differences are shown in Fig. 3. Comparing plasma at the center (A) and edge (B) from the actual cleaning process shows significant differences in plasma behavior between the two beam patterns. In the cleaning process using Line beam patterns, the increased heat input due to high laser beam overlap in the upper and lower edge areas caused more active evaporation and decomposition of paint in these areas compared to the center part. In the cleaning process using Square area beam patterns, uniform and stable plasma behavior is observed on the whole.

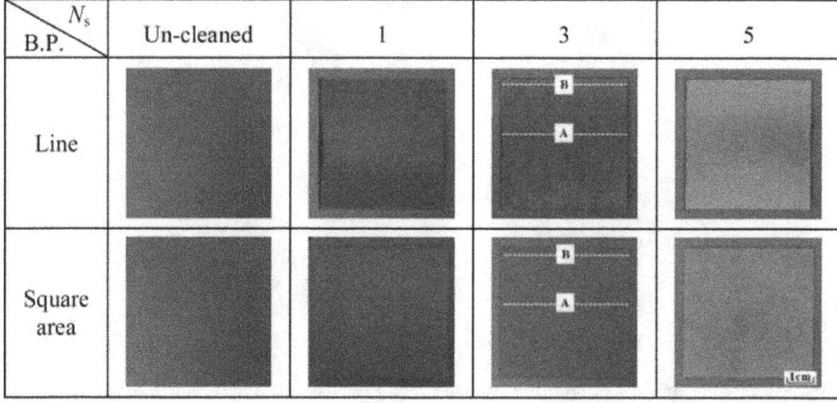

Fig. 2.   (Color online) Laser cleaned surfaces of anti-fouling painted steel with the laser beam pattern.

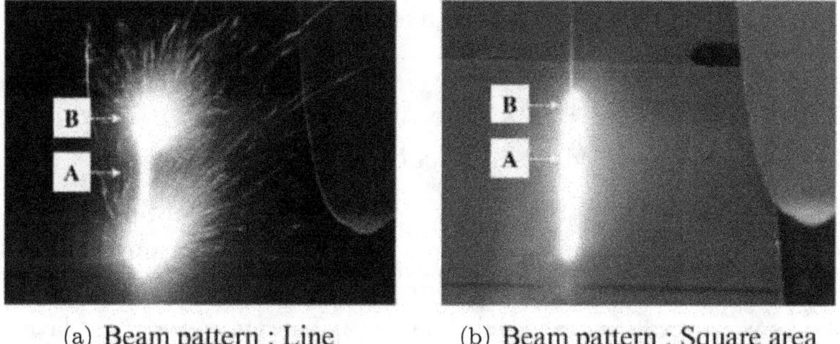

(a) Beam pattern : Line   (b) Beam pattern : Square area

Fig. 3.   (Color online) Comparison of plasma intensity in A and B zones according to beam pattern.

3.2. *Application of laser cleaning for oxide laser removal*

Laser cleaning is applied to the removal of the oxide layer on the steel surface using Square area beam patterns, which exhibits uniform laser cleaning performance, and the results are shown in Fig. 4. Prior to the application of laser cleaning, irregular red oxidized rust was formed on the specimen surface, which was completely removed after five scans of the laser. Table 2 shows the results of EDS spot analysis of the specimen before and after laser cleaning. Before laser cleaning, large amounts of O and Fe were detected on the specimen surface. This is because the main component of the oxide layer formed on the surface of specimen is iron oxide such as Fe_2O_3.

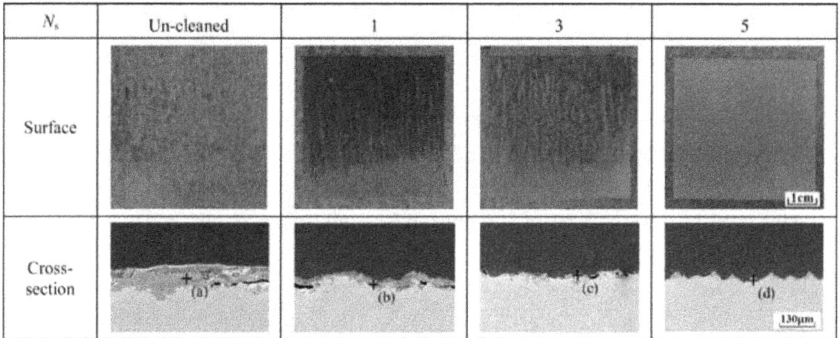

Fig. 4. (Color online) Laser cleaned surface and cross section of oxidized steel and EDS measurement positions.

Table 2. EDS spot analysis results on laser cleaned surfaces of oxidized steel (at.%).

		Laser cleaned		
Elements	(a) Uncleaned	(b) $N_s = 1$	(c) $N_s = 3$	(d) $N_s = 5$
O	48.02	45.36	43.55	3.66
Fe	51.98	54.64	56.45	96.34

The content of O and Fe was also high with one and three scans in which the oxide layer was not completely removed. On the other hand, with five scans in which the oxide layer was completely removed, a marked decrease in the content of O was confirmed. Therefore, these results confirm that laser cleaning allows for precise removal of the oxide layer as well as paint on the steel surface.

4. Conclusion

In this study, cleaning characteristics for paint and oxide layer removal of steel surfaces were investigated according to laser beam scanning methods. The results are as follows:

(1) In the result of laser cleaning experiment using Line beam patterns, nonuniform paint removal was confirmed in which paint was removed first in the edge area due to differences in the heat input depending on the area of the cleaning part according to the scanning method with stage movement.

(2) In the result of using Square area beam patterns with Galvano scanner controlled scanning, uniform laser cleaning performance was demonstrated throughout the cleaning area.

(3) This study demonstrated precise performance in the removal of paint and oxide layer on the steel surface with the application of laser cleaning technique, and this is expected to provide significant contribution to the development of eco-friendly surface cleaning method for building a smart yard.

References

1. J. E. Kim, M. S. Han and J. D. Kim, *Mod. Phys. Lett. B* **34**, 204042 (2020).
2. G. Zhang *et al.*, *Appl. Surf. Sci.* **506**, 144666 (2020).
3. G. Zhu *et al.*, *Opt. Laser Technol.* **132**, 106475 (2020).
4. J. E. Kim *et al.*, *J. Weld. Join.* **37**, 435 (2019).
5. J. E. Kim *et al.*, *J. Weld. Join.* **37**, 441 (2019).

Laser assisted synthesis of WS$_2$ nanorods by pulsed laser ablation in liquid environment*

Pankaj Koinkar[†], Kohei Sasaki[‡], Tetsuro Katayama[§] and Akihiro Furube[¶]

*Department of Optical Science, Tokushima University,
2-1 Minamijosanjima Cho, Tokushima 7708506, Japan*
[†] *koinkar@tokushima-u.ac.jp*
[‡] *c501938021@tokushima-u.ac.jp*
[§] *tetsuro@tokushima-u.ac.jp*
[¶] *furube.akihiro@tokushima-u.ac.jp*

Satoshi Sugano

*Tokushima University, 2-1 Minamijosanjima Cho,
Tokushima 7708506, Japan*
sugano@tokushima-u.ac.jp

Two dimensional (2D) materials are widely attracting the interest of researchers due to their unique crystal structure and diverse properties. In the present work, tungsten disulfide (WS$_2$) nanorods were synthesized by a simple method of pulsed laser ablation in liquid (PLAL) environment. The prepared WS$_2$ are analyzed by field emission scanning electron microscopy (FESEM), transmission electron microscopy (TEM), UV-visible spectroscopy (UV-vis) and Raman spectroscopy to confirm the surface morphology, phase and structure. A possible growth mechanism of WS$_2$ is proposed. This study indicates new door for the preparation of 2D materials with specific morphology.

Keywords: Laser ablation; WS$_2$ nanorods; UV-visible spectra; Raman spectra; growth mechanism.

1. Introduction

In recent years, many researches focused on the transition metal dichalcogenide (TMD) layered materials because of their distinctive layer structure

---

[†] Corresponding author.
*To cite this article, please refer to its earlier version published in the *International Journal of Modern Physics B*, Volume 35, 2140007 (2021), DOI: 10.1142/S0217979221400075.

as well as physical and chemical properties and they have been studied for extensive application.[1-3] Especially, uniform nanosized WS_2 nanostructures as compared to bulk counterparts have great potential applications in various fields such as energy harvesting, heterogenous catalysis and optoelectronics and so on.[4-6] In addition, as a result of quantum confinement effect, the indirect bandgap of WS_2 varies with its crystallinity, size and shape.[7-9] Till today, the WS_2 nanostructures with different surface morphologies have been prepared by employing several deposition techniques such as chemical vapor deposition, thermal decomposition, electrochemical exfoliation and hydrothermal method. These fabrication methods strongly depend on the use of high temperature environment, expensive gas precursors and therefore, the prepared nanostructure has been used for limited applications. It is equally important to understand that the role played by the vital factors such as larger bandgap with decreasing the number of layers, shape and size is useful for the fabrication of optoelectronic devices such as photodetector, field emission devices and so on. In this regard, pulsed laser ablation in liquid (PLAL) method is a simple, low cost, eco-friendly and relatively new method to prepare the various 2D nanomaterials under the family of TMD, which exhibits unique properties. In case of liquid exfoliation method like PLAL, the choice of solvent can significantly affect the final yield, the processing time and also the optical properties of the prepared nanostructures. WS_2 is considered as a hydrophobic material, so the choice of water as a solvent might require larger processing time as compared to other organic solvents. However, we have used deionized water as solvent to a record the UV-vis spectra in water medium which was one of the concern. In this study, we report on the synthesis of WS_2 nanorods prepared by PLAL method. The surface morphology, crystal structures and optical properties of prepared WS_2 nanorods were studied by SEM, TEM, UV-vis spectroscopy and Raman spectroscopy. Furthermore, a possible growth mechanism to reveal the formation of WS_2 is proposed.

2. Experimental Procedure

Synthesis of WS_2 nanosheet

The WS_2 powder used as a bulk material for the laser ablation was purchased from Sigma-Aldrich having the average particle size $\sim$1 to 2 μm. The ns laser ablation was performed using WS_2 powder (20 mg) in deionized water (20 mL). Before the laser ablation, the solution was under probe sonication for 10 min followed by ultrasonication for 20 min. Then solution was irradiated with constant pulses from the second harmonic generator of a Nd:YAG laser (wavelength 532 nm, repetition frequency 10 Hz, 10 ns fwhm/pulse, 55 mJ pulse^{-1}, Spectra Physics Quanta-Ray). The laser beam with a diameter of 5 mm was introduced into the solution and then was exposed horizontally for the laser ablation. The samples were treated for different irradiation times of 10 and 20 min. During the laser ablation, the solution was continuously stirred by a magnetic rotor. The surface morphology of the prepared samples was characterized by SEM (S4700, JEOL, Japan) operated at 5 kV and TEM (JEOL JEM-2100F, Japan) operated at 200 kV to identify the structural features and to confirm the formation of nanostructure. The UV-vis absorption spectra and Raman spectra were measured by using a V-670, JASCO spectrometer, and Nanophoton Raman-7 microscope to reveal optical features and chemical composition.

3. Results and Discussion

3.1. *UV-visible analysis*

UV-vis absorption spectra of bulk and laser ablated WS$_2$ samples are shown in Fig. 1. The increase in the absorption band can be seen for laser ablated WS$_2$ samples. It is seen that the WS$_2$ micro sheets are converted into nanorods and nanosheets after the laser ablation for 10 and 20 min, respectively. This change in absorption band suggests the formation of few layers of WS$_2$ separated from each other and then transformed into nanorods and nanosheets as compared to bulk WS$_2$ samples.

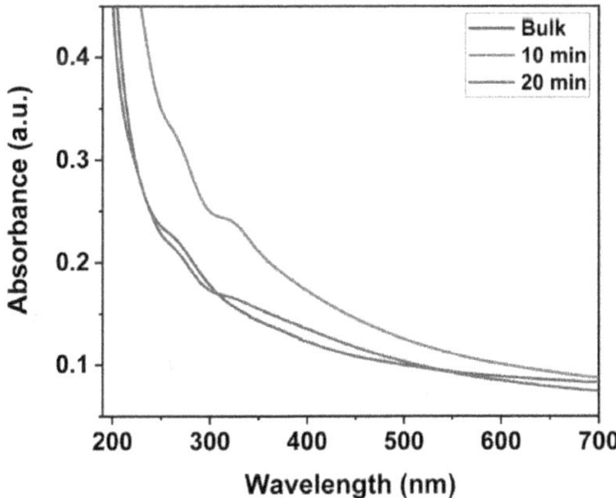

Fig. 1.   (Color online) UV spectra of (a) bulk WS$_2$ and (b, c) laser ablated WS$_2$ for 10 and 20 min.

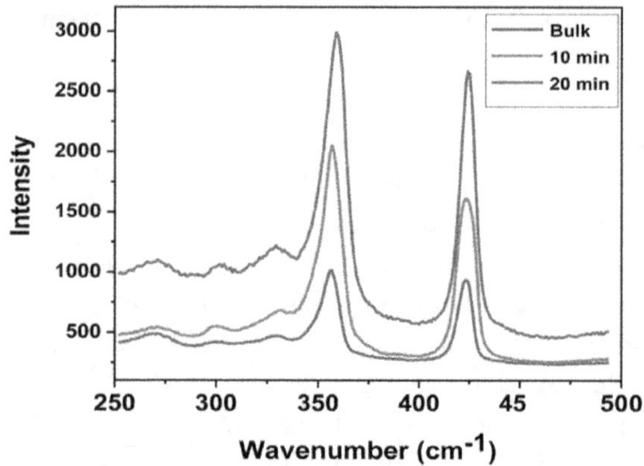

Fig. 2.   (Color online) Raman spectra of (a) bulk WS_2 and (b, c) laser ablated WS_2 for 10 and 20 min.

3.2. *Raman scattering spectra analysis*

Figure 2 depicts the Raman spectra of bulk, nanorods and nanosheets WS_2 recorded at room temperature. The spectra of three samples clearly show the presence of two characteristic active modes E_{2g}^1 and A_{1g}. For bulk WS_2, the characteristic mode appeared for E_{2g}^1 mode at 355.65 cm^{-1} is due to in-phase vibrations of W and S atoms and another characteristic A_{1g} mode appeared at 423.51 cm^{-1} belongs to the out-of-plane vibrations of the S atoms for the bulk WS_2. For WS_2 nanorods, both modes occurred at 356.63 and 423.13 cm^{-1}. It indicates the structure features of WS_2 nanorods is same as that of bulk WS_2. However, both modes appear at 358.65 and 424.12 cm^{-1} for nanosheets which suggests that the peak shift is in opposite direction as compared to an earlier report.[10]

3.3. *SEM and TEM analysis*

The surface morphology of bulk and laser ablated WS_2 sample was revealed by SEM as shown in Fig. 3. According to Fig. 3(a) of bulk WS_2, stacked structure of few layered sheets of WS_2 was observed. Figure 3(b) shows the formation of WS_2 nanorods having about 150 nm in diameter, a few microns in length, whereas, Fig. 3(c) shows WS_2 nanosheet with about 100 to 150 nm size. The TEM images shown in Fig. 4 clearly reveal the nanoparticle originated growth of WS_2 nanorods after the laser ablation for 10 min. Figure 4(a) confirms the presence of nanoparticles while Fig. 4(b) shows the formation of WS_2 nanorods.

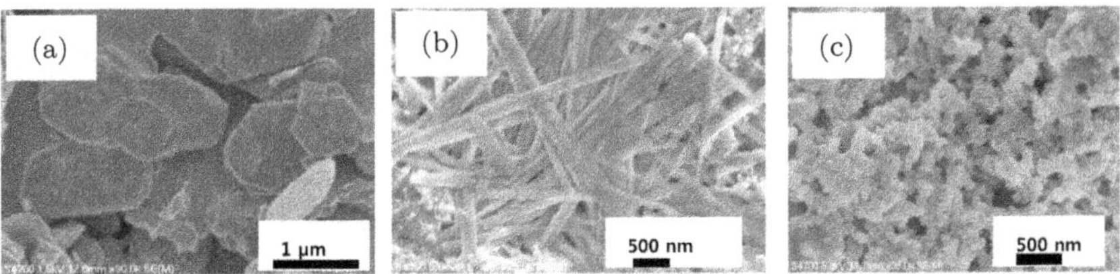

Fig. 3.   FESEM images of (a) bulk WS_2 and (b, c) laser ablated WS_2 for 10 and 20 min.

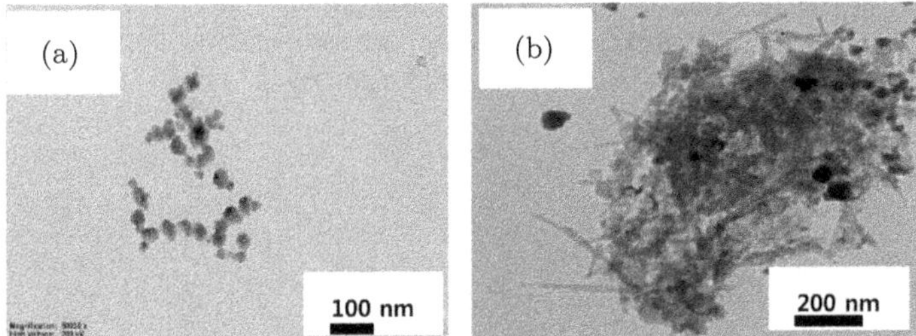

Fig. 4.   TEM images of (a, b) laser ablated WS_2 for 10 min at different regions.

3.4. *Growth mechanism of* WS_2 *nanorods*

Figure 4 shows the TEM image of long WS_2 nanorods formed by nanosecond laser treated for 10 min. The high pulse energy of nanosecond laser fragments the bulk WS_2 into small nanoparticles. Then these small nanoparticles can be used as a nucleation site to form the nanorods. It is believed the orientation of fragmented nanoparticles can play a vital role in transformation of nanorods.

4. Conclusion

We have the WS_2 nanorods by a pulsed laser ablation method. The result obtained from UV-visible spectra and Raman spectra indicated the formation of WS_2 nanorods and these WS_2 transformed into nanosheets/ nanoparticle with increase in the laser ablation time. In particular, this study reveals a new approach to grow WS_2 nanorods in liquid medium using nanosecond laser ablation for 10 min as compared to several methods which required more time. This synthesis route offers a potential to develop WS_2-based nanoelectronics and photonic devices.

References

1. R. Könenkamp, R. C. Word and C. Schlegel, *Appl. Phys. Lett.* **85**, 6004 (2004).
2. K. P. Musselman *et al.*, *Adv. Mater.* **22**, E254 (2010).
3. P. Koinkar, K. Sasaki and A. Furube, *Mod. Phys. Lett. B* **33**(14–15), 1940014 (2019).
4. R. Levi *et al.*, *Nano Lett.* **13**, 3736 (2013).
5. R. Morrish, T. Haak and C. A. Wolden, *Chem. Mater.* **26**, 3986 (2014).
6. S. Dhongade *et al.*, *ACS Appl. Nano Mater.* **3**(10), 9749 (2020).
7. G. Tang *et al.*, *Mater. Lett.* **65**, 3457 (2011).
8. Y. C. Wu *et al.*, *Mater. Lett.* **189**, 282 (2017).
9. T. W. Chen *et al.*, *Ultrason. Sonochem.* **56**, 430 (2019).
10. W. Ashraf *et al.*, *Appl. Nanosci.* **9**, 1515 (2019).

Development of multi-layered sewer pipe plug — 1st report: Ruptured test and stress analysis of protective sheet*

Nao-Aki Noda[†], Hisanori Tottori, Geng Gao, Rei Takaki and Yoshikazu Sano

Mechanical Engineering Department, Kyushu Institute of Technology, 1-1, Sensuicho, Tobata, Kitakyushu 804-8550, Japan
[†]*noda.naoaki844@mail.kyutech.jp*

Akira Kai

Hoshin Co., Ltd., 1-7-44, Aosaki, Oita 870-0278, Japan
kai@hoshin.co.jp

In recent years, the sewer system in Japan is becoming obsolete. It is, therefore, necessary to reinforce or repair without stopping sewer functions by applying a suitable water stopping method. In this study, a multi-layered sewer pipe plug consisting of a protective sheet and inner and outer rubber balls is focused since it can be installed and removed at the construction site in a short time conveniently. This sewer pipe plug has several advantages dealing with various diameters compared to the conventional type. In this study, the rupture test is conducted to improve the water stopping performance. The fractured position of the protective sheet of the sewer pipe plug is investigated experimentally. It is clarified that the maximum stress around the flange portion can be reduced by decreasing the flange inner diameter.

Keywords: Sewer pipe plug; protective sheet; ruptured test; stress analysis.

1. Introduction

Sewers occupy a considerable large amount of underground space compared to underground tunnels and subways. Japan's sewer system is becoming obsolete and has to be updated, reinforced and repaired after 2020.

[†]Corresponding author.
*To cite this article, please refer to its earlier version published in the *International Journal of Modern Physics B*, Volume 35, 2140015 (2021), DOI: 10.1142/S0217979221400154.

During the repair, it is necessary to apply a suitable water stopping method without stopping the sewer function. An air injection type pipe plug with one layer is used to stop water conventionally although the critical pressure is about critical pressure $p_{cr} = 0.3$ MPa or less with a risk of a rupture. Therefore, to stop high-pressure water for deep underground over 40 m, higher pressure resistance is required.

Figure 1 shows a three-layered sewer pipe plug to be developed.[1,2] The first inner layer is an inner ball made of natural rubber to keep airtight. The second layer is a protective sheet made by ultra-high molecular weight polyethylene fiber (Izanas fiber) to support the high internal pressure. The third outer layer is made of natural rubber to provide frictional resistance from the pipe wall. In this study, by performing a rupture test of the protective sheet, the strength and the fracture origin will be clarified. The stress analysis is also performed to improve the protective sheet performance.

2. Rupture Test for Protective Sheet Made by Ultra-High Molecular Weight Polyethylene Fibers

2.1. *Description of the protective sheet*

The protective sheet used in the pipe plug in Fig. 1 is made by the seamed cloth knitted by the thread composed of ultra-high molecular weight polyethylene fibers (UHMWPE). The UHMWPE fiber has a molecular weight of 15 times or more that of ordinary polyethylene having extremely high strength, excellent wear resistance, weather resistance and water resistance. This fiber has the highest level of strength and elastic modulus as an organic fiber. Table 1 shows the mechanical properties of UHMWPE fiber named IZANAS fiber. One thread is made by spinning 1170 UHMWPE fibers and the band cloth with 200 mm length and 100 mm width dimensions

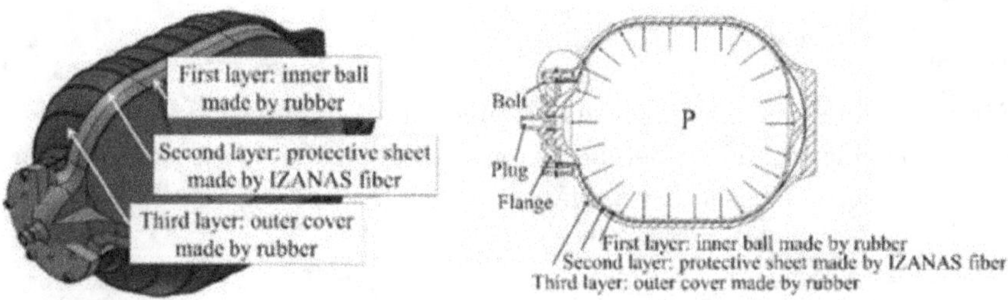

Fig. 1.   (Color online) Development of multi-layered sewer Pipe Plug.

Table 1. UHMWPE fiber named IZANAS used in the protective sheet.

Material	Diameter [μm]	Strength [MPa]	Elastic modulus [GPa]
Polyethylene	12	2.6–4.0×10^6	79

*One thread is made by spinning 1170 fibers.

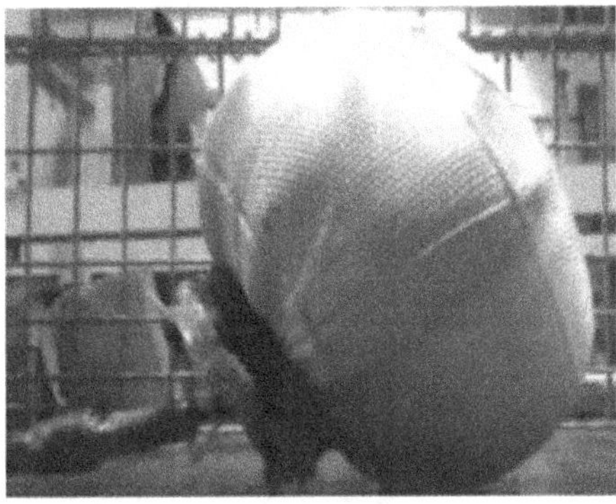

Fig. 2.   (Color online) Two-layered pipe plug used in the ruptured test.

is made by plain weaving the thread. Finally, the protective sheet is made
by sewing four-band clothes with Kepler thread composed of Aramid fibers.

2.2. *Experiment method*

Figure 2 illustrates the ruptured test of a two-layered sewer pipe plug consisting of an inner ball and a protective sheet. The strength of the protective
sheet is discussed identifying the fracture origin. Air is injected into the
inner rubber ball until the protective sheet ruptures. The fracture origin is
clarified by examining the protective sheet after it ruptured.

2.3. *Experimental results and discussion*

Figure 3 illustrates an example of a ruptured protective sheet. The rupture
occurs under the $p_{cr} = 0.53$ MPa, which is lower than the target pressure
$p = 2$ MPa to be used for the deep underground. Observation shows that
the protective sheet is fractured near the inner corner of the flange shown
in R_2 in Fig. 4. Note that the fracture origin is at the seamed portion near
the inner corner of the flange. Since the seamed portion is constrained in
deformation, the smaller deformation compared to the other portion may
increase the risk of fracture.

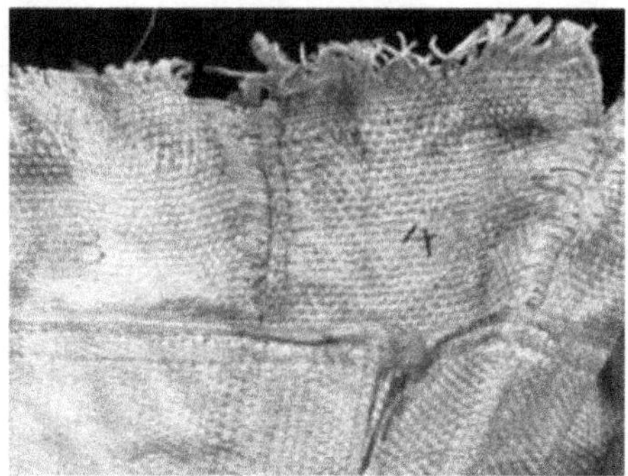

Fig. 3. Protective sheet after rupture.

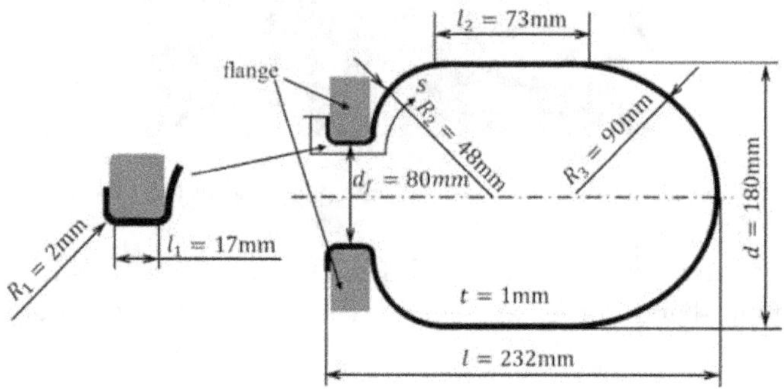

Fig. 4. FEM model for protective sheet when $d_f = 80$ mm.

3. Analytical Results and Discussion

3.1. *Analytical method*

Since the rigidity of the rubber ball is smaller and negligible compared to the protective sheet, only the protective sheet is analyzed by applying axisymmetric shell elements using general finite element method analysis software Marc/Mentat 2012. Figure 4 shows the analysis model. In Fig. 4, the circular arc R_1 is divided into 30 elements, R_2 is divided into 60 elements, R_3 is divided into 60 elements, straight line l_1 is divided into 34 elements and l_2 is divided into 146 elements. In this analysis, the flange inner diameter $d_f = 80$ mm, diameter $d = 180$ mm, thickness $t = 1$ mm, the elastic modulus $E = 3.4$ GPa, Poisson's ratio $\nu = 0.3$. The circumferential stress σ_θ, tangential stress σ_t and normal stress σ_n are obtained under the internal pressure $p = 1$ MPa.

3.2. *Analytical results and discussion*

Figures 5 and 6 show the analysis results for the flange inner diameters $d_f = 80$ mm and $d_f = 2$ mm. The solid line denotes the circumferential stress σ_θ, the dashed line denotes the tangential stress σ_t and the normal stress $\sigma_n = 0$. From Fig. 5, the maximum tensile stress $\sigma_t = 93$ MPa at the

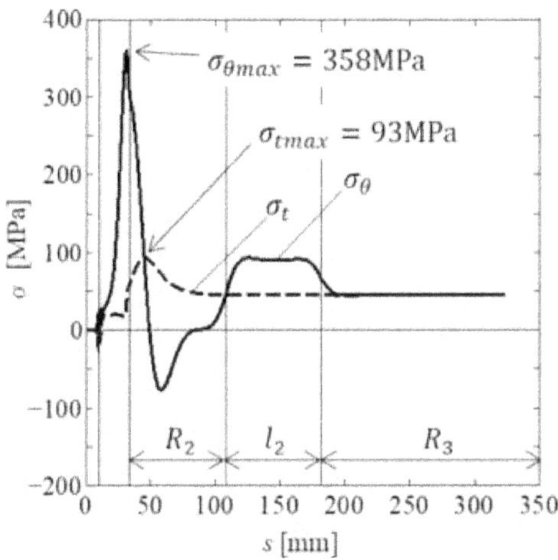

Fig. 5.   Stress distribution of protective sheet under $p = 1$ MPa when $d_f = 80$ mm.

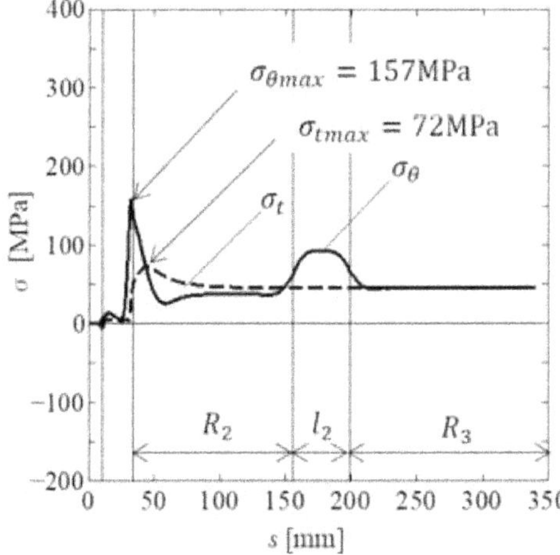

Fig. 6.   Stress distribution of protective sheet under $p = 1$ MPa when $d_f = 20$ mm.

seamed portion coincides with the fracture origin at the flange part in the ruptured test. Since the fracture occurred at an internal pressure of critical pressure $p_{cr} = 0.53$ MPa, the tensile strength of the seamed part of the protective sheet can be estimated as $\sigma_t = 93 \times 0.53 = 49$ MPa.

From the comparison of Figs. 5 and 6, it is seen that the maximum stresses σ_θ and σ_t around the flange portion decrease with decreasing the flange inner diameter d_f.

4. Conclusions

In this study, a rupture test was conducted for the multi-layered pipe plug to investigate the strength of the protective sheet. The stress analysis is also performed to clarify the fracture origin. The conclusions can be summarized in the following way.

(1) From the rupture test, it was found that the fracture originates from the seemed portion of the protective sheet is near the inner corner of the flange.
(2) From the stress analysis for the protective sheet, it was found that the maximum stress appears around the flange portion. The maximum stress can be estimated as $\sigma_t = 49$ MPa when fractured.
(3) It was clarified that the maximum stress around the flange portion can be reduced by decreasing the flange inner diameter.

References

1. R. Takaki *et al.*, *Proc. JSME* **No. 198-1**, G15 (2019).
2. G. Gao *et al.*, *Proc. JSME* **No. 208-1**, G26 (2020).

Development of multi-layered sewer pipe plug — 2nd report: Tensile strength of protective sheet bonded by seams*

Nao-Aki Noda,[†] Geng Gao, Hisanori Tottori, Rei Takaki and Yoshikazu Sano

*Mechanical Engineering Department, Kyushu Institute of Technology,
1-1, Sensuicho, Tobata, Kitakyushu 804-8550, Japan*
[†] *noda.naoaki844@mail.kyutech.jp*

Akira Kai

Hoshin Co., Ltd., 1-7-44, Aosaki, Oita 870-0278, Japan
kai@hoshin.co.jp

Since Japanese sewer system is becoming obsolete, it is necessary to reinforce and repair the system without stopping sewer functions. In this study, a multi-layered sewer pipe plug consisting of a protective sheet and inner and outer rubber balls is focused to be installed and removed conveniently at the construction site. In this paper, a suitable testing method and the strength of the protective sheet are discussed experimentally by changing the specimen geometry providing slit and seams. It is found that the slit specimen is most desirable to obtain the standard tensile strength of UHMWPE cloth named Izanas cloth. It is found that the seamed strength is $\sigma_B = 34$ MPa, which is about 17% of the standard tensile strength $\sigma_{B0} = 200$ MPa.

Keywords: Sewer pipe plug; protective sheet; UHMWPE fiber; bonded method.

1. Introduction

After 2020, most of Japanese sewerage system has to be updated and repaired because it is becoming obsolete. During the repair, it is necessary to develop a construction method without stopping the sewer function.

[†] Corresponding author.
*To cite this article, please refer to its earlier version published in the *International Journal of Modern Physics B*, Volume 35, 2140016 (2021), DOI: 10.1142/S0217979221400166.

Since the one-layered pipe plug conventionally used has a risk of burst, a three-layered pipe plug reinforced by the protective sheet should be proposed. In the preceding study, the rupture test showed that the fracture occurs at the seamed portion weaker than the other portion around the inner flange where the stress concentration appears. In this study, a tensile test will be conducted to clarify the strength of the protective sheet and the strength of the seamed portion.

2. Tensile Test of a Protective Sheet Made by Ultra-High Molecular Weight Polyethylene Fibers

The protective sheet used in the pipe plug is made by the seamed cloth knitted by the thread composed of ultra-high molecular weight polyethylene fibers (UHMWPE).[1–4] In this study, tensile tests are conducted to clarify the strength of the protective sheet itself and the strength of the seam part. Shimadzu universal testing machine AGS-J 10 kN is used as the tensile testing machine.

Figure 1 shows tensile specimens without a seam. Japan Industrial Standards L 1096 prescribes testing methods for woven and knitted fabrics stating that the size of the specimen length $l = 200$ mm, width $W = 50$ mm. In this study, $l = 100$ mm, $W = 50$ mm are used. The tensile speed of the test is 100 mm/min as prescribed in JIS L 1096. As shown in Figs. 1(b)

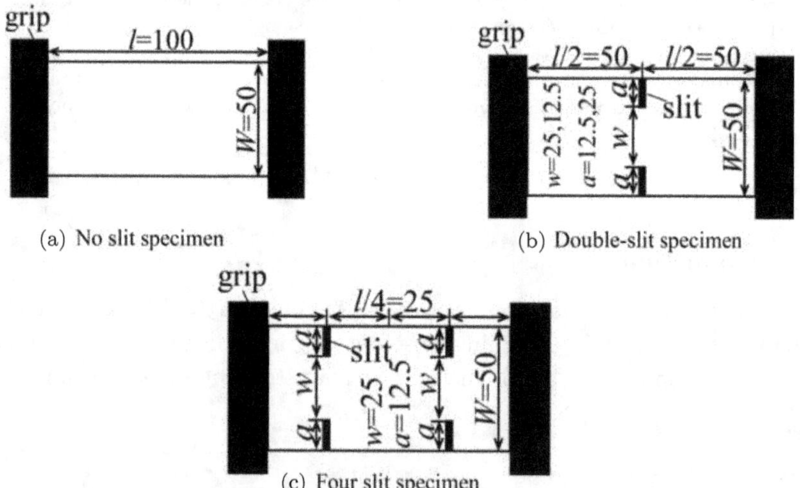

Fig. 1.   Tensile specimens whose minimum width $w = 50, 25, 12.5$ mm without seam.

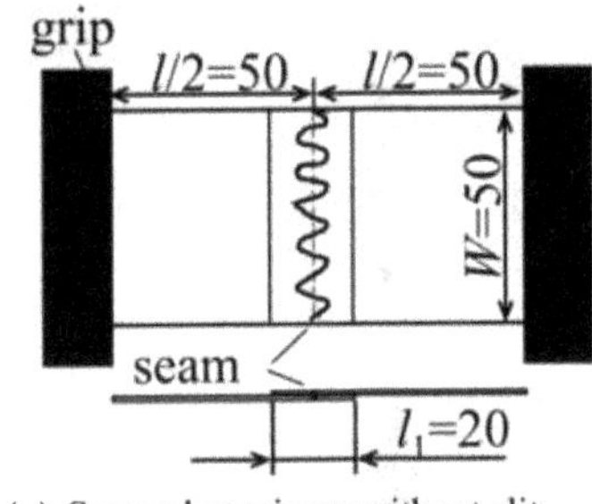

(a) Seamed specimen without slit

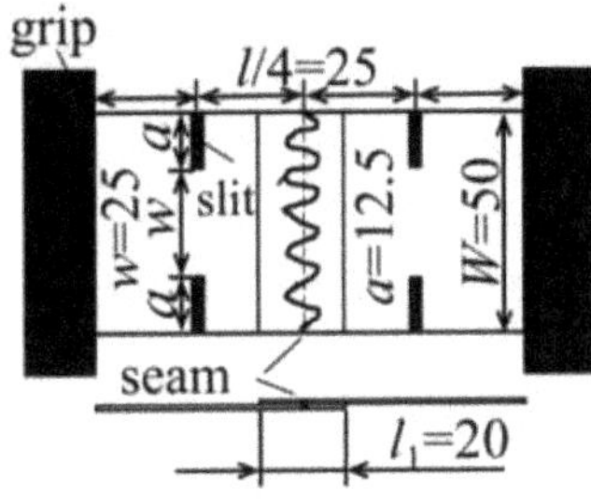

(b) Seamed specimen with four slits

Fig. 2.   Tensile specimens with seam.

and 1(c), the slit is considered to prevent the fracture at the griped portion although the slit is not prescribed in JIS. The specimens with slits have the minimum section width $w = 25$, 12.5 mm with thickness $t = 1$ mm. The griped portions at both ends of the test piece are reinforced by the adhesive and covered by gum tape. Figure 2 shows tensile specimens with seam whose overlap length $l_1 = 20$ mm is sewn at the center of the specimen.

3. Seamed Tensile Strength in Comparison with the Protective Sheet Strength

Figure 3 shows the stress–strain diagram for the specimen in Fig. 1(b). Here, the tensile strength σ_B is defined as the maximum value in Fig. 1(b). Table 1 summarizes the tensile strength obtained for the specimens in Fig. 1. The tensile strength σ_B in Table 1 is defined by the maximum tensile force P divided by the cross-sectional area $A = wt$. Here, the minimum width $w = 50$ mm for no slit and $w = 25$ mm, 12.5 mm for double slit and four slit specimens. The thickness of the specimen is always fixed as $t = 1$ mm.

In Table 1, the average tensile strength in Fig. 1(b) is defined as the standard tensile strength σ_{B0}. This is because the double-slit specimen in Fig. 1(b) is most suitable and the specimen without the slit in Fig. 1(a) is unsuitable because the fracture happened at the griped portion. However, by comparing the results of $w = 50$, 25, 12.5 mm, it is found that the tensile strength is nearly the same. It may be concluded that the effect of the minimum width w and the effect of the slit can be negligible.

Figure 3 shows the stress–strain diagram for the specimen without a seam in Fig. 1(a) compared to the specimen with a seam in Fig. 2(a). Table 2 summarizes the seamed tensile strength obtained for the seamed specimens

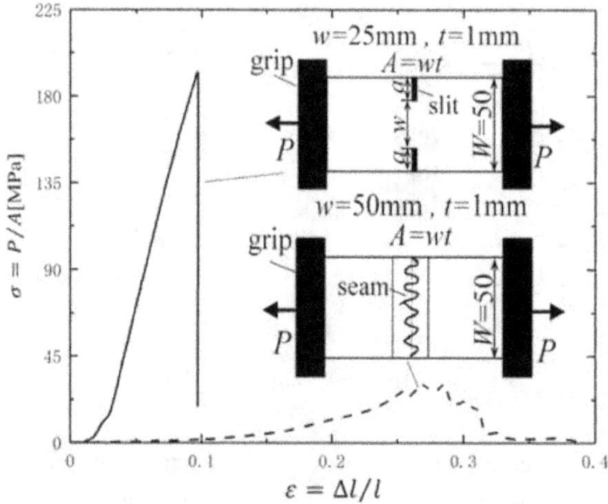

Fig. 3. Stress–strain curve for the specimen in Fig. 1(b) and the seamed specimen in Fig. 2(a).

Table 1. Tensile strength of UHMWPE cloth in Fig. 1.

Specimen	Sample No.	Number of slits	Minimum width w [mm]	Tensile strength [MPa] $\sigma_B = P/A$	Cross-sectional area A [mm^2]	Average tensile strength σ_B [MPa] (σ_B/σ_{B0})
Fig. 1(a)	A0-1	0	50	193	50×1	$\sigma_B = 195$ (98%)
	A0-2	0		197		
Fig. 1(b)	A1-1	2	25	201	25×1	Standard strength
	A1-2	2		195		$\sigma_{B0} = 200$
	A1-3	2	12.5	203	12.5×1	(100%)
	A1-4	2		201		
Fig. 1(c)	A2-1	4	25	178	25×1	$\sigma_B = 180$ (90%)
	A2-2	4		182		

Table 2. Seamed tensile strength of bonded UHMWPE cloth in Fig. 2.

Specimen	Sample No.	Number of slits	Minimum width w [mm]	Tensile strength [MPa] $\sigma_B = P/A$	Cross-sectional area A [mm^2]	Average tensile strength σ_B [MPa] (σ_B/σ_{B0})
Fig. 2(a)	B0-1	0	50	30	50×1	$\sigma_B = 34$ (17%)
	B0-2	0	50	38		
Fig. 2(b)	B1-1	4	25	34	25×1	$\sigma_B = 34$ (17%)
	B1-2	4	25	34		

in Fig. 2. From Tables 1 and 2, it is seen that the seamed strength $\sigma_B = 34$ MPa in Fig. 2(a) is 17% of the standard tensile strength $\sigma_{B0} = 200$ MPa in Fig. 1(b).

4. Conclusions

In this study, the tensile test is conducted to investigate the strength of the protective sheet made by UHMWPE cloth named Izanas cloth. The conclusions can be summarized in the following way.

(1) The effect of specimen geometry on the strength was invested. It is found that the slit specimen in Fig. 1(b) is most desirable to investigate the tensile strength of UHMWPE cloth named Izanas cloth. The specimen without the slit in Fig. 1(a) is not suitable because the fracture happened from the griped portion.

(2) The tensile strength of UHMWPE cloth named Izanas cloth can be estimated as $\sigma_{B0} = 200$ MPa. It is defined as the standard tensile strength of UHMWPE cloth named Izanas cloth.

(3) A tensile test is conducted to investigate the seamed strength of UHMWPE cloth named Izanas cloth in Fig. 2. It is found that the seamed strength is $\sigma_B = 34$ MPa, which is about 17% of the tensile strength of Izanas cloth $\sigma_{B0} = 200$ MPa.

References

1. R. Takaki *et al.*, *Proc. JSME* **No. 198-1**, G15 (2019).
2. G. Gao *et al.*, *Proc. JSME* **No. 208-1**, G26 (2020).
3. Japan industrial standards, JIS L 1096 Testing methods for woven and knitted fabrics.
4. N.-A. Noda *et al.*, *Int. J. Mod. Phys. B*, 2140015, doi:10.1142/502/7979224100154.

Effect of focal position on cut surface quality in laser cutting of 50-mm thick stainless steel*

Moo-Keun Song[†] and Jong-Do Kim[‡]

*Korea Maritime and Ocean University 727 Taejong-Ro, Yeongdo-Gu,
Busan 49112, South Korea*
[†]*mksong@kmou.ac.kr*
[‡]*jdkim@kmou.ac.kr*

Dong-Sig Shin[§], Su-Jin Lee[¶], Dae-Won Cho[‖]

*Busan Machinery Research Center,
Korea Institute of Machinery and Materials 48,
Mieumsandan 5-Ro 41beon-Gil, Gangseo-Gu,
Busan 46744, South Korea*
[§]*dsshin@kimm.re.kr*
[¶]*leesj@kimm.re.kr*
[‖]*dwcho@kimm.re.kr*

In this study, the parameters for underwater laser cutting of 50-mm thick stainless steel, which is typically used in nuclear power structures, are investigated. The focal position of laser beam significantly affects the cutting quality. In particular, in the cutting of the thick sample, change in the focal position determines the kerf width and the roughness of the cut surface. Moreover, the effects of the variation of kerf width and the cut surface characteristics on the focal position of the laser beam are investigated. As the focal position moved to the inside of the material, the upper kerf width increased, but the quality of the cut surface was improved.

Keywords: Laser cutting; stainless steel; focal position; cutting quality.

1. Introduction

High-power laser cutting that facilitates easy automation and long-distance remote control based on fiber transmission have attracted considerable

---

[‡]Corresponding author.
*To cite this article, please refer to its earlier version published in the *International Journal of Modern Physics B*, Volume 35, 2140018 (2021), DOI: 10.1142/S021797922140018X.

attention in the nuclear power plant demolition industry.[1-3] However, till now, the possibility of using laser cutting for thick materials has been the focus of this research, which showed satisfactory results for the metal dross produced during laser cutting. Nevertheless, because the metal dross can unnecessarily increase the waste volume when deposited on the wastes exposed to radiations, development of a technique that offers good cutting quality without metal dross production is important. To this effect, in this study, laser-cutting parameters have been investigated in terms of the structural parameters of 50-mm thick stainless steel, which is typically used in the nuclear power structures. The cutting quality can be considered as a parameter in the determination of the dross generation amount. Therefore, the changes in laser focal position, which can significantly affect laser-cutting quality of a thick plate, is primarily investigated among various parameters of laser cutting.

2. Experimental Methods

Stainless steel 316, typically used in real nuclear power structure, with a thickness of 50 mm was used in this study. A disk laser, which had 8 kW laser output and 200 μm beam diameter, was used for cutting. Nitrogen was used as the assist gas for cutting. Cutting was conducted using nitrogen as the assist gas, 1 mm as the stand-off distance, 20 bar as the gas pressure. The focal position, which is considered as a major parameter in this study, was changed at 5-mm intervals from 0 mm to 20 mm. Images of upper side, back side and cross-section of the laser-cut sample were captured, and the kerf width of the cut surface was measured. Furthermore, the drag line length was measured for the cut surface and the kerf width was measured by dividing the upper, middle and lower parts.

3. Results

3.1. *Laser-cutting parameter: focal position*

Figure 1 shows the upper side, back side, cut surface and cross-section of the laser-cut sample when the focal position was changed from 0 to $(-)20$ mm under the conditions of 8 kW laser output and 1.5 mm/s cutting speed. As the focal position moves from the surface toward the inside of material, the kerf width of the upper part increases. However, it is difficult to

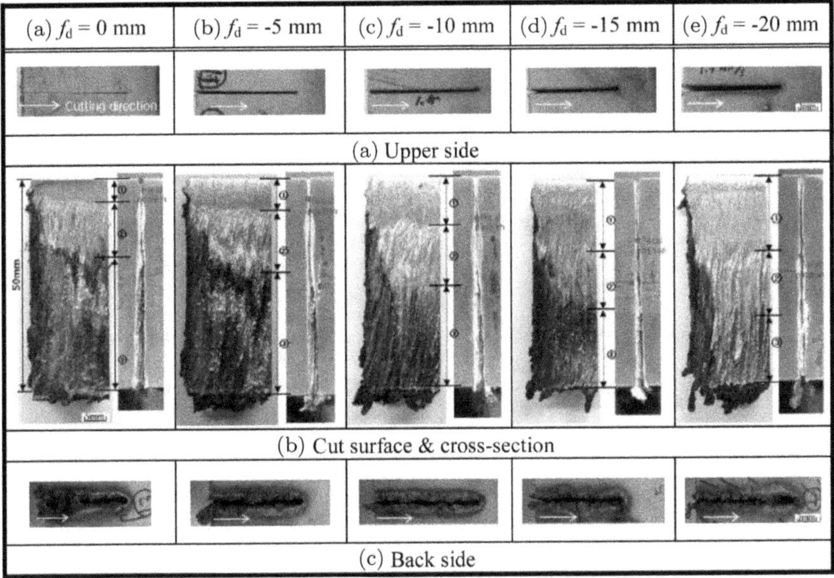

Fig. 1.   (Color online) Laser cut specimen shape according to focal position.

confirm whether the kerf width increases on the cut part of the back side because of dross generation. When the cut surface is examined, it can be seen that the drag line length of the upper part increases and the cutting quality improves as the focal position approaches $(-)20$ mm. At the drag line of the upper part, the material melts only because of high power laser energy during laser-cutting; thus, this part shows the best cutting quality. However, when the cut cross-section is examined, it is confirmed that the kerf width of the upper part gradually increases as the focal position moves deeper inside the material. This is because the diameter of the laser beam irradiated on the sample surface changes according to the change in the focal position. When the focal point is positioned at 0 mm, which is at the sample surface, the smallest focal point diameter is observed, and as the focal position moves toward the $(-)$ position, the focal point diameter on the sample surface increases. Therefore, if the focal position moves further toward the $(-)$ side surpassing $(-)20$ mm, the energy density of laser beam on the surface will considerably increase, thereby hindering laser cutting.

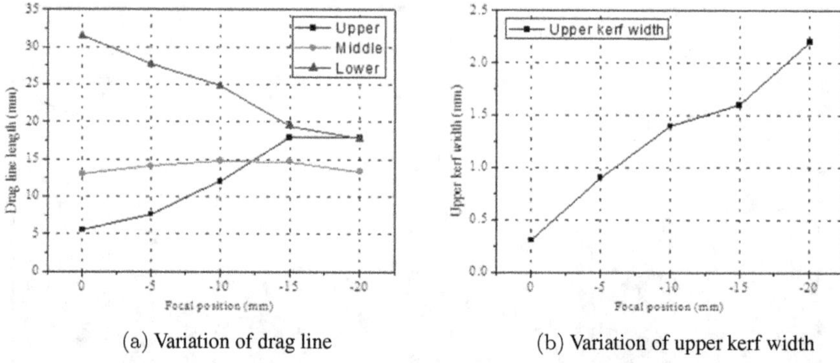

(a) Variation of drag line   (b) Variation of upper kerf width

Fig. 2.   (Color online) Variation of drag line and kerf width.

Figure 2 shows the changes in kerf width and drag line length of the upper part of the cut surface as a function of focal position. As explained above, when the focal position moves to the ($-$) side, the kerf width and drag line of the upper part increase, while the drag line of the lower part decreases. Therefore, it can be seen that when the focal position of the laser beam is at ($-$)20 mm, the kerf width of the upper part increases and the overall cutting quality is good because of the increase in the drag line of the upper part.

3.2. *Microstructural characteristics of cut section*

In the cut cross-sections of the above-mentioned figures, areas shown in bright colors around the kerf width can be seen. To investigate the structural characteristics of this area, the microstructure of the cut section was examined. For example, Fig. 3 shows the microstructure of a cut sample under the conditions of ($-$)20 mm focal. The structure of the center part shows that after melting, the recast metal forms a layer at both flank sides of the cut section and shows a solidified shape. On the other hand, the lower part shows that the melted metal has a solidified shape without being discharged in the cut section. This solidified melted metal has a role in decreasing the cut surface quality and improving the roughness.

Figure 4 shows the energy dispersive spectrometry (EDS) component analysis results for the dross in the undischarged recast zone and lower part. The melted part and the dross did not show much difference from the elemental composition of other basic material areas. However, the oxygen

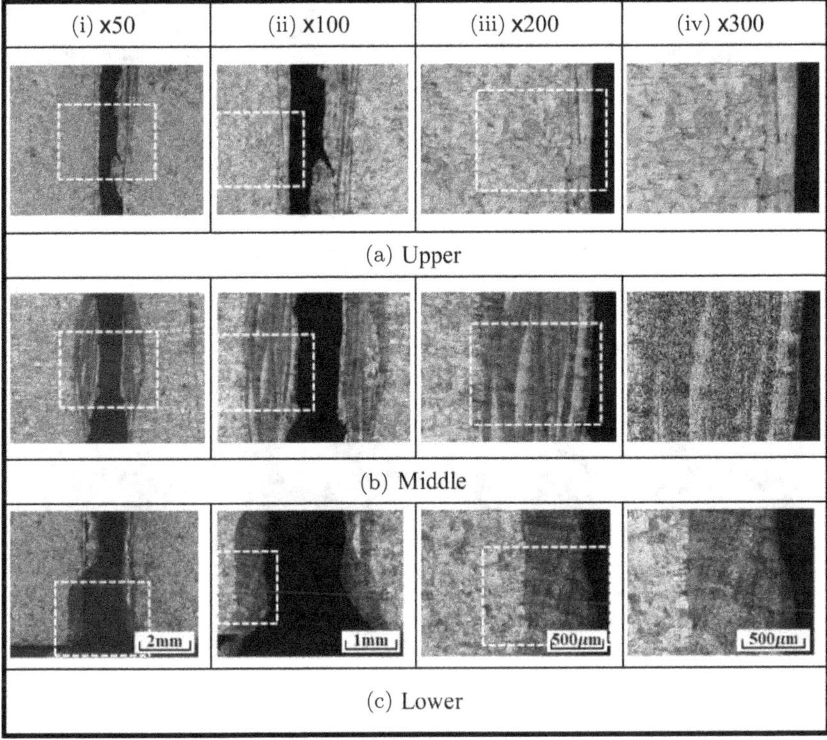

Fig. 3. (Color online) Microstructure of laser cut specimen.

distribution of dross was higher than that of other areas. This implies that the melted part at the center was blocked from contacting with oxygen due to the presence of nitrogen, which was the cutting gas. In contrast, dross was generated as the melted metal reacted with surrounding oxygen during the solidifying process after being discharged.

This study conducted an experiment by changing the laser focal position in a fixed state of laser output to derive the parameters that minimize dross generation, which was inevitable. It was confirmed that the change in the focal position was a major parameter in the determination of the kerf width and the cut surface quality.

Furthermore, the optimal condition was derived for the best cutting quality in this study. In future, a condition that can inhibit dross generation will be investigated through the optimization of other parameters such as gas pressure and diameter and shape of the cutting nozzle.

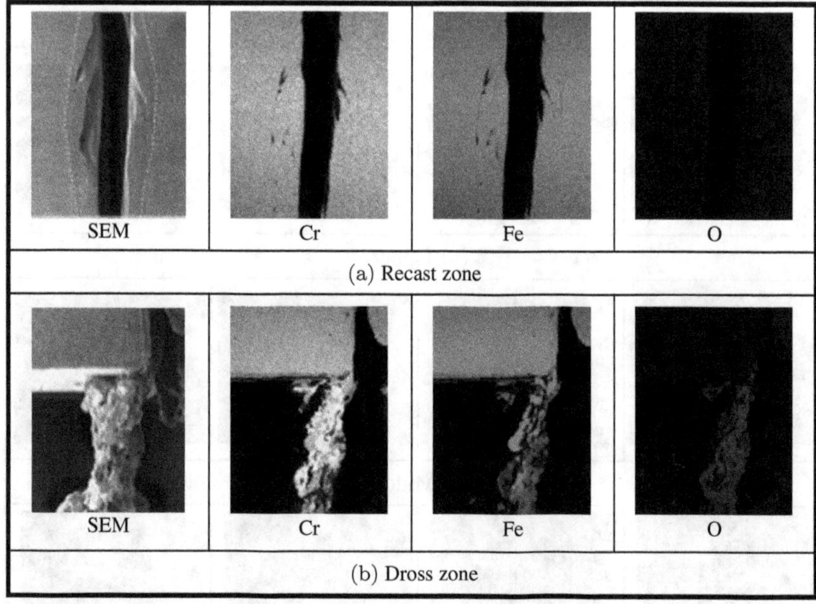

Fig. 4.   (Color online) EDS analysis results.

4. Conclusions

In order to inhibit dross generation during laser-cutting of a thick stainless steel plate, the cutting parameters based on the laser beam's focal position were investigated, and the following results were obtained.

(1) On changing the focal position from 0 to (−)20 mm, the upper kerf width increased as the focal position moved toward the (−) side. Moreover, because the upper drag line increased, the overall cutting quality improved.
(2) In the microstructure of the cut section, the undischarged melted part existed in the cut surface; although this melted part showed almost same elemental composition as that of the base material, oxygen was detected in the dross part.
(3) Although dross generation could not be inhibited, the results confirmed that the change in the focal position was a major parameter for determining the quality of laser-cut surface.

References

1. S. Stelzer *et al.*, *Phys. Procedia* **41**, 299 (2013).
2. A. B. Lopez *et al.*, *Nucl. Eng. Technol.* **49**, 865 (2017).
3. D. W. Cho *et al.*, *J. KWJS* **38**, 569 (2020).

Stress due to interfacial slip causing sleeve fracture in shrink-fitted work roll*

Nao-Aki Noda[†], Rahimah Abdul Rafar[‡] and Yoshikazu Sano

*Department of Mechanical Engineering,
Kyushu Institute of Technology, 1-1 Sensui-cho,
Tobata-ku, Kitakyushu-shi 804-8550, Japan*
[†] *noda.naoaki844@mail.kyutech.jp*
[‡] *eimah7178@gmail.com*

The rolls are classified into two types; one is a single-solid type, and the other is a shrink-fitted construction type consisting of a sleeve and a shaft. The bimetallic work rolls are widely used in the roughing stands of hot rolling stand mills. Regarding a shrink-fitted construction type, the interfacial slip sometimes appears between the shaft and the shrink-fitted sleeve. This interfacial slip can be regarded as the relative displacement between the sleeve and the shaft. In this paper, the stress due to the interfacial slip is studied because the stress may cause the sleeve fracture. It is found that the stress in the shrink-fitted surface is slightly decreased with increasing number of rotations n. Therefore, the stress obtained by the simulation at $n = 2$ can be used to estimate the fatigue strength.

Keywords: Shrink-fitting; rolling roll; bimetallic roll; interfacial slip; motor torque; sleeve; shaft.

1. Introduction

Rolling rolls are essential to the iron and steel industries, and therefore lots of efforts have been done to improve their mechanical properties. Figure 1 illustrates the rolling roll in roughing stands of hot rolling stand mills. The rolls are classified into two types; one is a single-solid type, and the other is a shrink-fitted assembled type consisting of a sleeve and a shaft. In shrink-fitted sleeve roll, although the wear and surface roughness soon

---

[‡]Corresponding author.
*To cite this article, please refer to its earlier version published in the *International Journal of Modern Physics B*, Volume 35, 2140020 (2021), DOI: 10.1142/S0217979221400208.

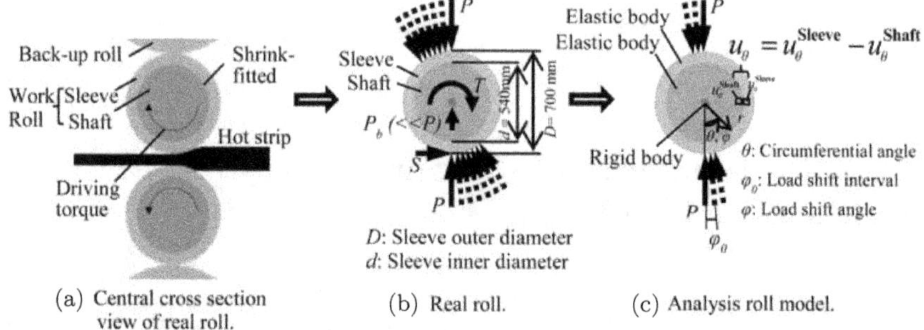

Fig. 1. Modeling and simulation for interfacial slip when P is the load from back-up roll and hot-strip, S is the frictional force, P_b is the bending force from bearing, T is the driving torque.

appear on the roll body, the shaft can be reused. However, it is clarified in the previous study that this shrink-fitted structure has generated the interfacial slip between the sleeve and the shaft.[1-3] Furthermore, in order to consider the roll breakage caused by the stress due to the interfacial slip, the stress state in the shrink-fitted surface will be focused in this paper.

2. Numerical Simulation Method of Interface Slip Under the Action of Shaft Drive Torque

Figure 1 illustrates two-dimensional modeling in numerical simulation. Figure 1(a) illustrates the central cross-section of real roll and Fig. 1(b) illustrates the real roll by shifting the load on the roll surface with the roll center fixed. Here, the motor torque T is balanced with the frictional moment S as $T = SD/2$ where D is the sleeve outer diameter. Figure 1(c) shows the analysis roll model that is used in the study. In Fig. 1, the roll is subjected to compressive force P from the back-up roll, the rolling reaction force P_h, the frictional force S from the strip, the bending force P_b from the bearing and torque T from the motor. Since two-dimensional model is applied, the external force per unit length should be considered. The loading condition used in this study is the concentrated loading P from the back-up roll and the reaction P from the strip with $P = 13{,}270$ N/mm. Here, when the rated torque of the motor is $T_m = 471$ Nm/mm, the frictional force can be calculated as $S = 1346$ Nm/mm. The shrink-fitting ratio is defined as δ/d, where δ is the diameter difference between the inner diameter of the sleeve and d is the sleeve inner diameter. In this study, $\delta/d = 0.5 \times 10^{-3}$ is used with the friction coefficient $\mu = 0.3$ between the sleeve and the shaft.

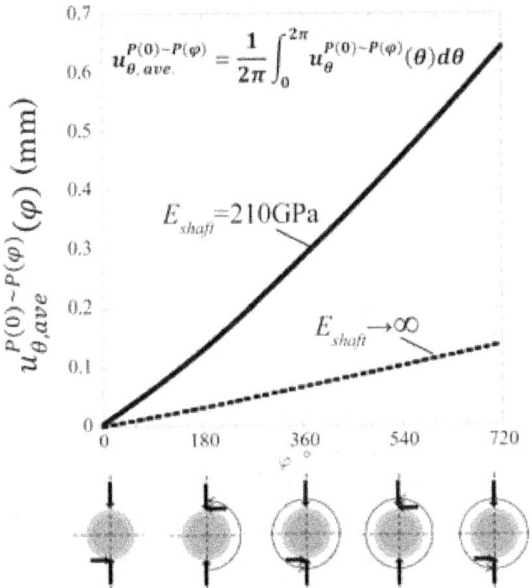

Fig. 2.   Interfacial displacement for shaft Young's modulus $E_{\text{shaft}} = 210$ GPa in comparison with $E_{\text{shaft}} \rightarrow \infty$.

In this analysis, the circumferential relative displacement at the interface between the sleeve and shaft when load move from the angle $\varphi = 0°$ to $\varphi = \varphi$ is defined as $u_\theta^{P(0)\sim P(\varphi)}$. Here, notation φ denotes the load shifting angle and θ denotes the position where the displacement is evaluated. Thus, the accumulation of the displacement is given by Eq. (1).

$$u_{\theta,\text{ave.}}^{P(0)\sim P(\varphi)} = \frac{1}{2\pi}\int_0^{2\pi} u_\theta^{P(0)\sim P(\varphi)}(\theta)d\theta. \tag{1}$$

Figure 2 shows the interfacial displacement $u_{\theta,\text{ave.}}^{P(0)\sim P(\varphi)}$ by rotating the load twice for shaft Young's modulus $E_{\text{shaft}} \rightarrow \infty$ and $E_{\text{shaft}} = 210$ GPa. From the figure, the amount of interfacial slip $u_{\theta,\text{ave.}}^{P(0)\sim P(\varphi)}$ increases almost linearly with increasing the load shift angle φ. The interface slip may occur as soon as the load shifting starts.

3. Stress on the Inner Surface of the Sleeve

In the sleeve assembly type roll, it should be noted that the circumferential slippage sometimes occurs even though the resistance torque at the interface is larger than the motor torque. Due to this slip, a localized adhesion occurs and causes a crack initiation. Then, the crack propagates and forms a groove-like flaw. The repeated stress concentration at this flaw increases the flaw dimension and finally cause sleeve fracture.[4] In this way, the stress σ_θ appearing at the shrink-fitted surface should be estimated since σ_θ is the largest stress component and causes such damage.

Figure 3 shows the stress distribution $\sigma_{\theta,T}^{P(0)\sim P(2\pi)}$, $\sigma_{\theta,T}^{P(0)\sim P(4\pi)}$ and $\sigma_{\theta,T}^{P(0)\sim P(6\pi)}$ when the number of rotations $n = 1$, 2 and 3, respectively, for (a) $E_{\text{shaft}} \rightarrow \infty$ and (b) $E_{\text{shaft}} = 210$ GPa. In Fig. 3(a), there is almost no difference in the stress distribution σ_θ when $n = 1$, 2 and 3. From Fig. 3(a), the maximum stress is $\sigma_{\theta,\max} = 121$ MPa and the minimum stress is $\sigma_{\theta,\min} = 72$ MPa. In Fig. 3(b), the interfacial stress obtained shows differences in the stress distribution when $n = 1$, 2 and 3. From Fig. 3(b), the maximum stress is $\sigma_{\theta,\max} = 171$ MPa and the minimum stress is $\sigma_{\theta,\min} = 16$ MPa. Regarding the possibility of fatigue fracture from the inner surface of the sleeve due to these stresses, it is known that the inevitable tensile residual stress has a large effect on the inner surface of the sleeve.

The stress amplitude at which failure occurs for a given number of rotations is simplified in Fig. 4. In Fig. 4, stress amplitude σ_a and mean stress σ_m for $E_{\text{shaft}} \rightarrow \infty$ shows almost constant values as n increases. As for $E_{\text{shaft}} = 210$ GPa, the fatigue strength shows the safety variation

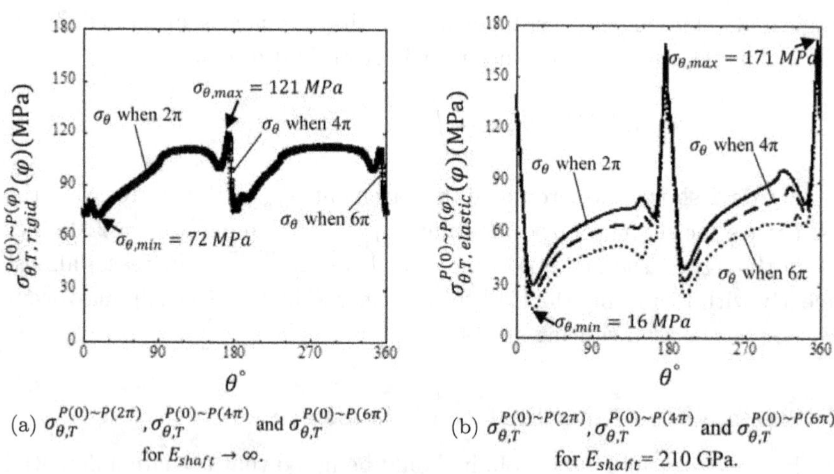

(a) $\sigma_{\theta,T}^{P(0)\sim P(2\pi)}$, $\sigma_{\theta,T}^{P(0)\sim P(4\pi)}$ and $\sigma_{\theta,T}^{P(0)\sim P(6\pi)}$ for $E_{shaft} \rightarrow \infty$.

(b) $\sigma_{\theta,T}^{P(0)\sim P(2\pi)}$, $\sigma_{\theta,T}^{P(0)\sim P(4\pi)}$ and $\sigma_{\theta,T}^{P(0)\sim P(6\pi)}$ for $E_{shaft} = 210$ GPa.

Fig. 3. $\sigma_{\theta,T}^{P(0)\sim P(2\pi)}$, $\sigma_{\theta,T}^{P(0)\sim P(4\pi)}$ and $\sigma_{\theta,T}^{P(0)\sim P(6\pi)}$ when number of rotations $n = 1$, 2 and 3 of $E_{\text{shaft}} \rightarrow \infty$ and $E_{\text{shaft}} = 210$ GPa.

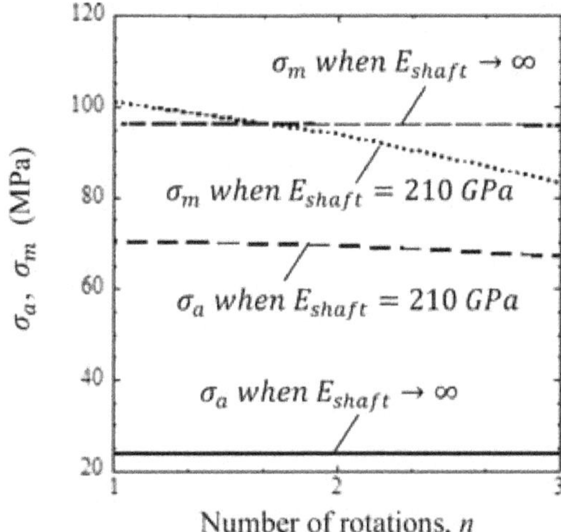

Fig. 4. Stress amplitude σ_a and mean stress σ_m for $E_{\text{shaft}} = 210$ GPa in comparison with $E_{\text{shaft}} \to \infty$.

because stress amplitude σ_a and mean stress σ_m decreasing as the number of rotations increases.

4. Conclusion

In this paper, the stress due to the interfacial slip is studied because the stress may cause the sleeve fracture. In this study, the circumferential stress σ_θ should be estimated since σ_θ is the largest stress component and may cause sleeve damage. From the analysis, it was found that the stress in the shrink-fitted surface for $E_{\text{shaft}} \to \infty$ shows almost constant values as the number of rotations n increases. As for $E_{\text{shaft}} = 210$ GPa, the stress in the shrink-fitted surface is slightly decreased with increasing n. Therefore, the stress obtained by simulation at $n = 2$ can be used to estimate the fatigue strength.

References

1. H. Sakai *et al.*, *Tetsu-to-Hagane* **105**, 1126 (2019).
2. H. Sakai *et al.*, *Tetsu-to-Hagane* **105**, 411 (2019).
3. H. Sakai *et al.*, *ISIJ Int.* **59**, 889 (2019).
4. E. Matsunaga, T. Tsuyuki and Y. Sano, *CAMP-ISIJ* **11**, 362 (1998).

Experimental prediction of internal defects according to defect area on NDI via water absorption behavior*

Kyo-Moon Lee, Soo-Jeong Park, Tianyu Yu, Seong-Jae Park and Yun-Hae Kim[†]

*Major of Materials Engineering,
Department of Marine Equipment Engineering,
Korea Maritime and Ocean University, 727 Taejong-ro,
Yeongdo-gu, Busan 49112, Republic of Korea*
[†] *yunheak@kmou.ac.kr*

This study analyzed the relationship between the defect area identified through a C-scan and the void volume in CF-PEKK composite materials through the water absorption behavior to predict the void volume. The water absorption content varies with the defect area; however, the defect area identified through a C-scan and the water absorption content did not show a proportional relationship. This is because voids are distributed in the through-thickness. The results indicated that the absorption behavior could be used to predict the void volume. Irreversible absorption was found to be independent of the void volume. Further, no matrix degradation was seen with water immersion at 70°C; however, some local swelling was seen.

Keywords: Carbon-PEKK composites; water absorption; void volume; C-scan.

1. Introduction

Thermoplastic composite materials show improved mechanical performance, such as impact resistance and abrasion resistance, as well as recyclability. However, the thermoplastic forming process is complicated. Studies of the out-of-autoclave process technology are still underway.[1,2] A thermoplastic composite laminate is mainly manufactured by stacking prepregs. When using this process for manufacturing large-scale products, defects such as splicing, tow gap and overlap can occur repeatedly, thereby causing

---

[†]Corresponding author.
*To cite this article, please refer to its earlier version published in the *International Journal of Modern Physics B*, Volume 35, 2140021 (2021), DOI: 10.1142/S021797922140021X.

internal defects in the laminate and degrading its mechanical properties.[3,4] Nondestructive inspection (NDI) methods like A- and C-scans can determine the defect position. However, they provide only a qualitative value, and it remains difficult to quantitatively predict the void volume of a laminate and to determine the void position. Polyetherketoneketone (PEKK) composite materials generally have high resistance to water absorption and are rarely affected by high-temperature moisture, thereby suffering little degradation of the resin and their mechanical properties.[5,6] However, if such composite materials have a void defect, water can fill this void defect and thus increase the absorption saturation.[7] In this light, the present study aims to determine the relationship between the defect area identified through a C-scan and the internal void defects in a carbon-fiber-reinforced PEKK laminate (CF-PEKK) and to predict the volume of internal void defects from the water absorption behavior.

2. Experimental Works

The CF-PEKK composite consists of unidirectional carbon-fiber-reinforced PEKK prepregs from Solvay (APC PEKK-FC). The matrix content has $\sim$34 wt.%, and it is a semicrystalline thermoplastic. It has a glass transition temperature (T_g) of 159°C and a melting temperature (T_m) of 337°C. The composite laminate was manufactured using a vacuum-bag-only process at 380°C for 1 h and with $[45/0/\text{-}45/90]_{4s}$ orientation. Then, as shown in Table 1, specimens were classified according to the defect area identified through a C-scan.

A water absorption test of CF-PEKK was conducted according to ASTM D5229. CF-PEKK was immersed in distilled water at 70°C.

Table 1.   Characteristic of the specimens.

Type	Nondefect		Half-Defect		Full-Defect	
Defect	Free-defect		$50 < \% \leq 80$		$80 < \% \leq 100$	
Area (%)	0	0	68	62	100	99
Max/Min (dB)	91.5/88.1	91.5/88.1	83.0/50.0	83.0/50.0	50/40.0	70.0/40.0
NDI Image						

The specimen weight was measured periodically using a scale with 0.1-mg accuracy until the specimen reached saturation. To observe the effect of water immersion, the saturated specimens were dried at 70°C for 48 h and weighed to calculated the water absorption content (irreversible absorption). Then, the defects identified through the C-scan were observed using an optical microscope, and the degradation of the CF-PEKK was observed by scanning electron microscopy (SEM).

3. Results and Discussions

Figure 1(a) shows a graph of the water absorption content of CF-PEKK with the different defect areas. The water absorption content (M) is calculated as

$$M(\%) = \frac{(W_f - W_i)}{W_i} \times 100, \tag{1}$$

where W_f and W_i are the weight after time and initial oven-dry weight, respectively. Absorption occurs quickly within 24 h. Half-defect (HD) and full-defect (FD) specimens contain internal defects such as voids, debonding and delamination and thus differ significantly from non-defect (ND) specimens. Although the FD specimen clearly shows a different defect area and damping factor relative to the HD specimen, the water absorption content of the HD specimen lies within the deviation of the water absorption content of the FD specimen. It indicates that the defects in the HD specimen are distributed in the through-thickness (Fig. 2).

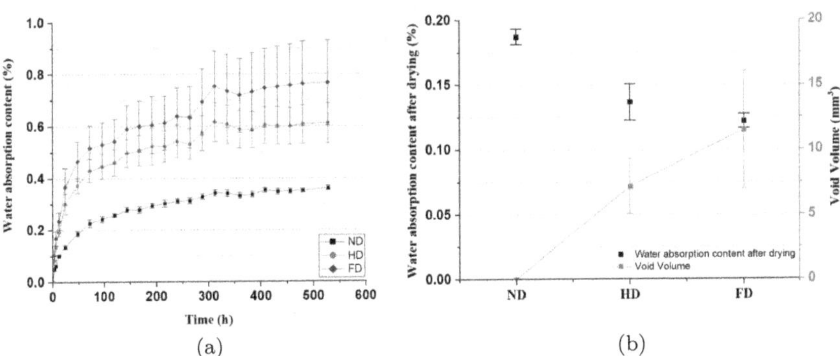

Fig. 1. (Color online) Water absorption content of CF-PEKK (a) and water absorption after drying (b) depending on defect areas.

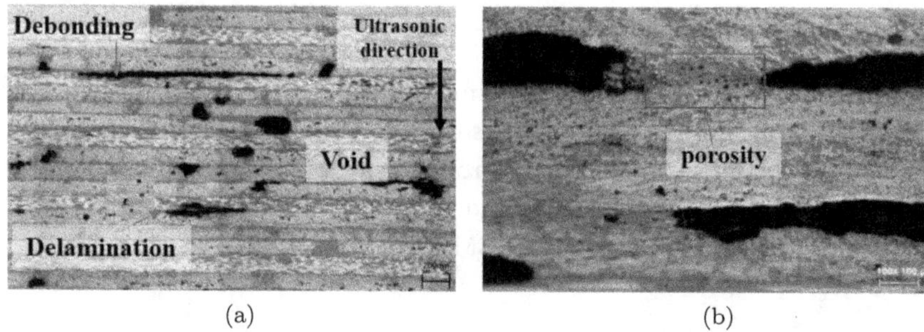

Fig. 2. The optical microscope observation to specify defect of CF-PEKK on NDI: (a) ×50, (b) ×100.

Figure 1(b) shows the void volume and irreversible absorption. The void volume of the specimen was calculated as

$$\text{Void volume (mm}^3) = (M_{\text{satu}} - M_{\text{refer}})\frac{W_i}{D_\circ C_{\text{water}}} \times 0.01, \qquad (2)$$

where M_{satu}, M_{refer} and $D_\circ C_{\text{water}}$ are the saturation content of water absorption, saturation content of water absorption of ND specimen and density of water at 70°C. In particular, M_{refer} is the water absorption content of the ND specimen. Equation (2) assumes that there is no degradation owing to water absorption. The irreversible absorption was calculated using Eq. (1) above. The void volume of HD and FD specimens is 7.19 mm^3 and 11.53 mm^3, respectively. The ND specimen showed a high irreversible absorption of 0.18%, whereas the HD and FD specimens showed a low irreversible absorption of 0.14% and 0.12%, respectively. The irreversible absorption decreased with increasing void volume; however, in the FD specimen, irreversible absorption showed a very small deviation whereas the void volume showed a large deviation. Therefore, irreversible absorption is concluded to be independent of the void volume.

Figure 2 shows defects such as voids, debonding, delamination and porosity in the defect area identified through the C-scan. Many defects also exist in the form of vertical overlaps [Fig. 2(a)]. The distribution of voids in the defect area is not identified through the C-scan.

Figure 3 shows SEM images of the ND, HD and FD specimen surfaces that were obtained to determine the degradation of CF-PEKK in high-temperature water. A comparison of the surface before immersion [Fig. 3(a)] and after immersion [Fig. 3(b)] revealed no matrix degradation.

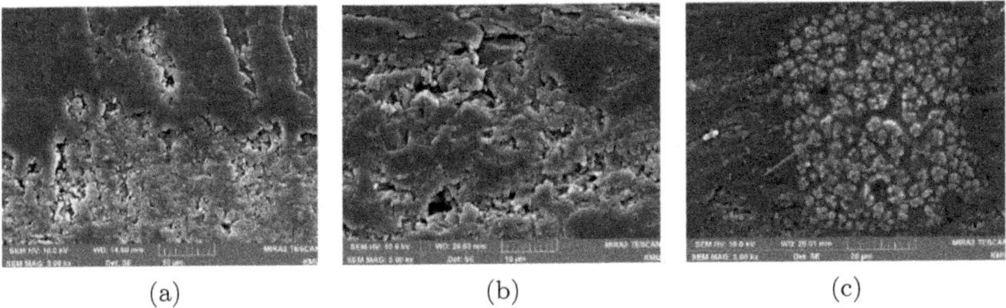

(a)   (b)   (c)

Fig. 3. SEM analysis of matrix surface before immersion (a), after drying water (b) and swelling trace (c).

However, after immersion, all specimen surfaces showed traces of local swelling [Fig. 3(c)].

4. Conclusion

The absorption behavior varies significantly depending on the presence of internal voids. Therefore, it can potentially be used to predict the void volume. By contrast, the defect area identified through NDI shows limited potential to predict the void volume because voids are distributed in the through-thickness. Further, irreversible absorption is independent of the void volume. Finally, high-temperature water immersion does not affect the degradation of CF-PEKK, although it induces local swelling. A future study will aim to clarify the relationship between defects and irreversible absorption and the effect of swelling on the void volume.

Acknowledgments

This work was supported by the Technology Innovation Program (No. 20007444) funded By the Ministry of Trade, Industry & Energy (MOTIE, Korea).

References

1. J. U. Jang *et al.*, *Funct. Compos. Struct.* **2**, 045008 (2020).
2. H. Jung *et al.*, *Funct. Compos. Struct.* **2**, 025001 (2020).
3. M. Lan *et al.*, *Compos. A: Appl. Sci. Manuf.* **82**, 198 (2016).
4. W. Woigk *et al.*, *Compos. Struct.* **201**, 1004 (2018).
5. T. Juska, *J. Thermoplast. Compos. Mater.* **6**, 256 (1993).
6. R. L. Mazur *et al.*, *J. Reinf. Plast. Compos.* **33**, 749 (2014).
7. M. L. Costa *et al.*, *Polymer Plast Tech Eng.* **45**, 691 (2006).

Photographic image processing to predict radiation dermatitis in breast cancer patients using machine learning algorithms*

Chou-Hsien Lee[†,‡], Chen-Lin Kang[§], Chin-Dar Tseng[†], Chi-Ming Chou[†],
Chin-Shiuh Shieh[†], Chih-Hsueh Lin[†], I-Hsing Tsai[†], Bo-Sheng Li[†],
Jia-Hong Ren[†], Pei-Ju Chao[†,§] and Tsair-Fwu Lee[†,¶,‖]

[†]*Medical Physics and Informatics Laboratory of Electronics Engineering,
National Kaohsiung University of Science and Technology,
Kaohsiung 80778, Taiwan, ROC*

[‡]*Department of Radiation Oncology, E-Da Hospital,
Kaohsiung 82445, Taiwan, ROC*

[§]*Department of Radiation Oncology,
Kaohsiung Chang Gung Memorial Hospital
and Chang Gung University, College of Medicine,
Kaohsiung 83342, Taiwan, ROC*

[¶]*PhD program in Biomedical Engineering,
Kaohsiung Medical University, Kaohsiung 80708, Taiwan*
[‖]*tflee@nkust.edu.tw*

Radiation therapy is an essential part of the comprehensive breast cancer treatment strategy, and radiation dermatitis is the inevitable side-effect. According to either patient-related or treatment-related factors, patients will experience different degrees of acute radiation dermatitis. This study proposes a machine learning architecture based on image and time series features. Using the skin image of the irradiated part during radiotherapy, the image feature is extracted with a gray-level co-occurrence matrix (GLCM) and color space, combined with the time series feature with gradient boosting decision trees (GBDT) to predict the severity of dermatitis after seven days of treatment. The results show that, through the combination of image and time series features, the predicted accuracy (ACC) and area under the curve (AUC) can be effectively improved to 0.8 and 0.85 respectively. The results of GBDT show higher prediction accuracy and robustness than AdaBoost algorithm. This framework can be used as

---

[‖]Corresponding author.
*To cite this article, please refer to its earlier version published in the *International Journal of Modern Physics B*, Volume 35, 2140022 (2021), DOI: 10.1142/S0217979221400221.

an auxiliary diagnostic tool to assist doctors in making appropriate treatments before severe dermatitis occurs, in order to reduce the radiotoxicity caused by radiotherapy of patients.

Keywords: Machine learning; radiation dermatitis; breast cancer; gradient boosting decision trees.

1. Introduction

Breast cancer is one of the most common cancer types in women worldwide, and radiotherapy (RT) plays an important role in the multi-modalities treatment strategy, in both lumpectomy and mastectomy settings. While improving local control and overall survival, however, breast radiation therapy has radiation dermatitis as an inevitable side-effect. How to minimize the skin toxicity without jeopardizing the RT plan is a paramount issue. Several managements such as using topical corticosteroid and RT techniques with IMRT have proved to be somewhat effective in the amelioration of breast radiation dermatitis.[1] Although radiation dermatitis develops with a predictable timing pattern, there is individual variation due to either patient-related or treatment-related factors. A reliable prediction model is needed to aid clinical professionals in providing better health care for breast cancer patients during the course of RT.

The research on image processing technology and machine learning in the prediction of RT toxicity is gradually increasing.[2-4] It is a common practice to use a gray-level co-occurrence matrix (GLCM) as an image processing technique to extract texture and color features.[5] These features are more robust to the image angle and illumination differences after being standardized. Gradient boosting decision trees (GBDT) use a boosting method (Boosting) to minimize the loss function,[6] so that the model has good prediction accuracy, and provides a reliable and effective machine learning computing.[7]

2. Materials and Methods

The dataset is taken from a prospective trial assessing the effectiveness of No Sting Barrier Film that enrolled breast cancer patients undergoing RT during 2016–2017 at a single institute. The Institutional Review Board (IRB) number is EMRP29104N. The data set contains the image records of 62 de-identified patients who were followed for 25 or 30 days during the course of breast cancer RT. Each patient has 5 to 9 images of the affected part of the breast with different treatment periods. These images have corresponding radiodermatitis levels as image labels. These labels were given by a radiation oncologist based on the five classification standards for radiation dermatitis by the Radiation Therapy Oncology Group (RTOG) of the United States. In the breast cancer images of the 62 patients, the numbers of level 0, 1 and 2 images were 93, 153 and 146 respectively, and the number of level 3 and 4 images was 0. Level 0 and level 1 images were classified as "slight", and level 2 images were classified as "severe".

The classifier uses a machine learning method called GBDT, which minimizes the loss function by iteratively generating regression trees, and maps the outputs of all regression trees to the corresponding. The category label output by the GBDT model can be obtained to predict the severity of dermatitis (mild or severe) of the image in the next week. The experimental process is shown in Fig. 1.

In order to predict the severity of dermatitis caused by RT in breast cancer patients, the experiment uses a GLCM and four color spaces to extract image texture features and color features. These image features were merged with the time series feature. Random forest (RF) can filter out some important features. The detailed process of feature extraction is shown in Fig. 2.

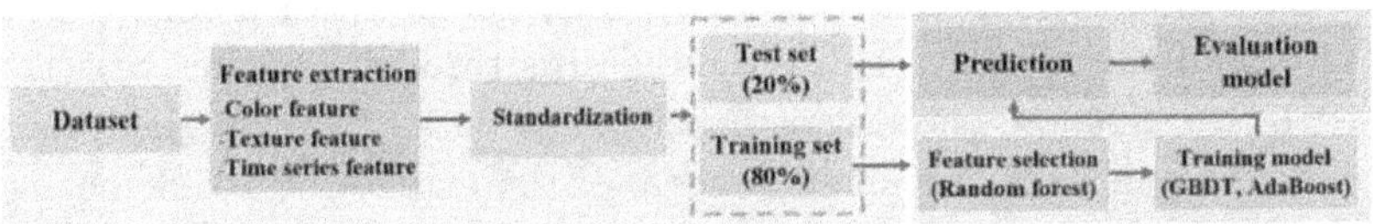

Fig. 1. (Color online) The experimental process flowchart.

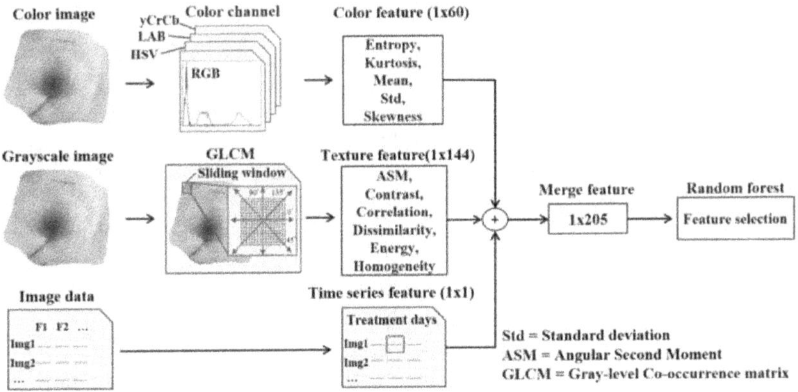

Fig. 2. (Color online) Feature extraction flowchart.

The color image in Fig. 2 is converted into four standard color spaces, RGB (red, green, blub), HSV (hue, haturation, value), LAB (Lab color space) and YCrCb (luminance, chrominance). Each color space has three channels. Therefore, by calculating five statistics of all channels, 60-dimensional features will be generated. The GLCM extracts 144-dimensional features through grayscale images, which are obtained through six types of statistics of six distances (1, 3, 5, ..., 12) and four angles (0, 45, 90, 135 degrees). The time series feature is the number of days of treatment for each image.

3. Results and Discussion

AdaBoost was used as a benchmark for method comparison. The model evaluation indicators use accuracy (ACC) and the area under the curve (AUC). The experimental results are shown in Table 1. The ACC and AUC predicted by GBDT using image features are 0.703 and 0.660, respectively. After adding the time series feature, the ACC and AUC rise to 0.805 and 0.851, respectively. Compared with the AdaBoost method, the ACC and AUC of GBDT are higher by 0.023 and 0.038. The above results indicate that machine learning can obtain the relationship between time characteristics and the severity of future dermatitis through training, and then assist in learning the characteristics of image data. Through this predictive model, clinicians can know in advance the extent of possible changes in the patient's subsequent development of radiation dermatitis. Furthermore, providing appropriate health education and preventive medical care will help improve the overall severity of radiation dermatitis.

Table 1.   The experimental results.

Algorithm	Feature	Evaluation	Mean	Standard Deviation
AdaBoost	Image feature	ACC	0.679	0.042
		AUC	0.644	0.063
	Image feature +	ACC	0.782	0.043
	Time series feature	AUC	0.813	0.059
GBDT	Image feature	ACC	0.703	0.027
		AUC	0.660	0.055
	Image feature +	ACC	0.805	0.041
	Time series feature	AUC	0.851	0.043

ACC: accuracy; AUC: area under the curve; GBDT: gradient boosting decision trees.

4. Conclusions

This study focuses on the combination of image feature and time series feature extraction technology and machine learning. The results confirm that the proposed framework has a certain accuracy in predicting the severity of future radiation dermatitis. This verifies the feasibility and potential of the combination of image and time series features. As an auxiliary diagnostic system, this architecture can help physicians to find the possibility of acute dermatitis early, and provide patients with prospective medical treatment.

Acknowledgment

This work was supported by the grant No. MOST 109-2221-E-992-011-MY2.

References

1. F. Haruna *et al.*, *Anticancer Res.* **37**, 5343 (2017).
2. K. Saednia *et al.*, *Int. J. Radiat. Oncol. Biol. Phys.* **106**, 1071 (2020).
3. L. J. Isaksson *et al.*, *Front. Oncol.* **10**, 790 (2020).
4. O. Maillot *et al.*, *Cancer/Radiothér.* **22**, 205 (2018).
5. R. B. Oliveira *et al.*, *Neural Comput. Appl.* **29**, 613 (2018).
6. J. H. Friedman, *Ann. Statistics* **29**(5), 1189 (2001).
7. A. K. Verma *et al.*, *Asian Pacific J. Cancer Prev.* **20**, 1887 (2019).

Mechanical properties of large-scale composite structures according to process conditions*

Chang Wook Park[†] and Sung Won Yoon[‡]

*Research Institute of Medium & Small Shipbuilding,
38-6, Noksansandan 232, Kangseo-gu, Busan 46757, Republic of Korea*
[†] *cwpark@rims.re.kr*
[‡] *swyoon@rims.re.kr*

The purpose of this study is to determine the correct estimation of the concept design and laminating pattern for composites. In this study, the effect of carbon tow prepreg on the composite structure has been investigated. Also, this work evaluated the structural safety according to the laminating pattern and process technique by computer simulation. The laminating pattern of three kinds was designed through the normal shape. Also, process technique of two kinds were carried out in dry and wet method to produce composites structure. The result of FEA showed that the structural safety of carbon tow prepreg is not significantly superior to that of the structure with wet method material. However, considering the benefits of working environment and the reduction of the number of process, the possibility of replacing materials and process technique will be sufficient for composite structure.

Keywords: Carbon fiber reinforced plastic; finite element analysis; composite super structure; carbon tow preg; filament winding.

1. Introduction

In shipbuilding industry area, the application of composites has increased for weight reduction of hull and environmental protection. In lightweight materials, composites with excellent specific strength and specific stiffness are emerging as next-generation materials. In particular, Carbon Fiber Reinforced Plastic (CFRP) has characteristics of high strength, high elasticity and high corrosion resistance. So, it is a suitable material for reducing

---

[‡]Corresponding author.
*To cite this article, please refer to its earlier version published in the *International Journal of Modern Physics B*, Volume 35, 2140025 (2021), DOI: 10.1142/S0217979221400257.

the weight of the ship components such as super structure. In addition, applying the composites has the advantage of reducing vibration, reducing component count and increasing the efficiency of energy. Therefore, demand for the development of equipment that possesses characteristics such as strength, weight reduction continues to increasing.[1–3]

2. Design of the Composites Structure

2.1. *Material properties*

The mechanical properties of CFRP were evaluated for winding pattern analysis of a composites structure. In this study, specimens were prepared by applying pretension load using filament winding method and cured in an oven at 80°C. The mechanical properties of specimens were measured by tensile, compression and shear strength tests using a universal test machine according to the ASTM standards. The mechanical properties of the CFRP laminates for simulation are shown in Table 1.[4]

2.2. *Finite element analysis*

In the finite element model of composite materials, it is very important to confirm that the winding patterns and stacking sequence of reinforcements is correctly set. In this study, a simulation was carried out to improve the structural safety for the load transferred to the structure through various angles and laminate designs. Therefore, the winding patterns of reinforcement was applied at ±30, ±45 and ±60 degrees. In addition, the lamination of the reinforcement in the composite materials are a symmetrical type in general. Thus, the laminate stacking sequence was defined as a symmetry form. The meshing controls for FEA has various types of elements

Table 1.   Mechanical properties of the CFRP.

Properties	Wet	Dry	Properties	Wet	Dry
Tensile strength X [MPa]	1630	2130	Compressive strength Z [MPa]	65	102
Tensile strength Y [MPa]	32	30	Shear strength XY [MPa]	80	60
Tensile strength Z [MPa]	32	30	Density [kg m^{-3}]	1508	1490
Compressive strength X [MPa]	700	1000	Poisson's Ratio	0.27	0.27
Compressive strength Y [MPa]	65	102			

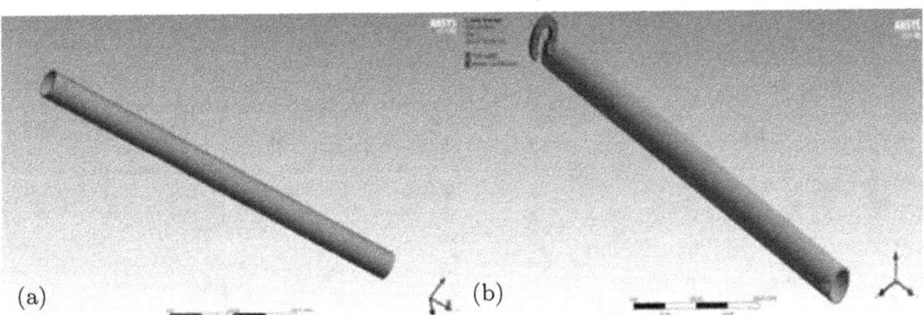

Fig. 1. (Color online) (a) Shape of Mesh for composite structure (b) loading and boundary conditions.

depending on the mechanical behavior required for the structure.[5,6] Figure 1 show the loading and boundary conditions for intermediate shaft model.

3. Results and Discussions

For comparison with the composite structure according to winding patterns, the composite structure applied to the two kinds of carbon fiber used for filament winding process were evaluated under the same conditions. The static structure results of the FEA of the dry winding composite structure in the direction of 30° show that stress concentration around the end of the structure occurred at the maximum stress of 374.16 MPa. Also, the static structure result of the FEA of the dry winding composite structure in the direction of 45° was confirmed with the maximum stress of 337.79 MPa. The result of the direction of 60° showed a stress of 362.84 MPa. On the other hand, the static structure results of the FEA of the wet winding composite structure in the direction of 30° show that stress concentration around the end of the structure occurred at the maximum stress of 373.59 MPa. Also, the static structure result of the FEA of the dry winding composite structure in the direction of 45° was confirmed with the maximum stress of 338.09 MPa. The result of the direction of 60° showed a stress of 362.33 MPa. Normal stress showed the lowest value in the 30° direction, and Y-axis and Z-axis were identified as the lowest stress value at 60°. The IRF showed the most stable tendency in a 45° direction. From the point of view of the Dry fiber and Wet fiber, which are applied materials for composite structure, stress showed no significant difference. However, the IRF value in the direction of 45° showed better value than the dry, which is judged by the effect of material strength on shear stress. It is confirmed

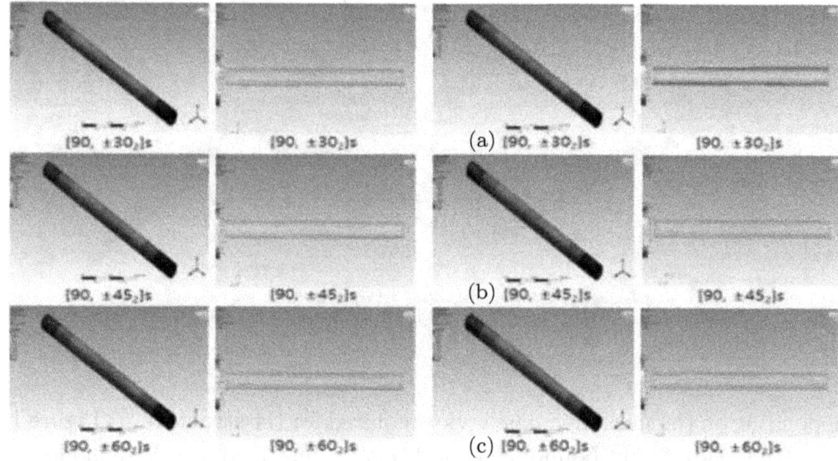

Fig. 2.   (Color online) Result of the static structure.

that this is due to the difference in physical properties between dry winding and wet winding, and it is considered that the strength characteristic of the material influences stress concentration. However, it was confirmed that the stress concentration did not reach the breaking strength of the applied material. Therefore, it is not influenced by structural safety.

Determination of the stacking angles is the most important factor in the design of structures for composite materials, and the expression of performance may depend on the stacking angle, even for structures of the same thickness. In order to determine the stacking angle of the propulsion shaft of the composite material, the mobility of the propulsion shaft must be reviewed and the direction of stress transferred must be analyzed to ensure the safety of the structure. The change in elasticity coefficient compared to the stacking angle was confirmed by the following Eqs. (1)–(3).

$$\frac{1}{E_{x\text{lamina}}} = \frac{1}{E_{11}}C^4 + \left[\frac{1}{G_{12}} - \frac{2\nu_{12}}{E_{11}}\right]S^2C^2 + \frac{1}{E_{22}}S^4, \tag{1}$$

$$\frac{1}{E_{y\text{lamina}}} = \frac{1}{E_{11}}S^4 + \left[\frac{1}{G_{12}} - \frac{2\nu_{12}}{E_{11}}\right]S^2C^2 + \frac{1}{E_{22}}C^4, \tag{2}$$

$$\frac{1}{G_{xy\text{lamina}}} = 2\left[\frac{2}{E_{11}} - \frac{2}{E_{22}} + \frac{2\nu_{12}}{E_{11}} + \frac{1}{G_{12}}\right]S^2C^2 + \frac{1}{G_{12}}\lfloor C^4 + S^4\rfloor. \tag{3}$$

— Exlamina : Elastic modulus of lamina in X-direction, MPa
— Eylamina : Elastic modulus of lamina in Y-direction, MPa

— E11 : Longitudinal elastic modulus of lamina, MPa
— E22 : Transverse elastic modulus of lamina, MPa
— G12 : Shear modulus of lamina in 12-direction, MPa
— V12 : Major Poisson's ratio - C : $\cos\theta$/S : $\sin\theta$

4. Conclusions

In this study, the possibility of the application of composite materials for a composite structure is analyzed by FEA and the following conclusion was drawn. The structural safety of the composite structure with dry winding method is not significantly superior to that of the composite structure with wet winding method. However, considering the benefits of weight reduction and the reduction of the number of components, the possibility of replacing composite structure with dry winding method will be sufficient.

Acknowledgments

This work was supported by the Technology Innovation Program (20010950, Development of Large-Scale Equipment for Eco-Ships) funded by the Ministry of Trade, Industry & Energy (MOTIE, Korea).

References

1. H. J. Hwang *et al.*, *Funct. Compos. Struct.* **2**, 035003 (2020).
2. S. Karimi, A. Salamat and S. Javadpour, *J. Mech. Sci. Tech.* **30**, 1755 (2016).
3. A. Alikhani *et al.*, *Funct. Compos. Struct.* **2**, 035005 (2020).
4. S. Kyriakides *et al.*, *Int. J. Solids Struct.* **32** 689 (1995).
5. S. W. Tsai, *Introduction to Composite Materials* (Routledge, Abingdon-on-Thames, 1980).
6. S. T. Pinho, L. Iannucci and P. Robinson, *Compos. A Appl. Sci. Manuf.* **37** 63 (2006).

How to estimate fatigue strength of wood material*

Kinjirou Saitou and Nao-Aki Noda[†]

*Department of Mechanical Engineering,
Kyushu Institute of Technology Laboratory,
Kitakyushu-Shi, Fukuoka 804-0015, Japan*
[†]*noda.naoaki844@mail.kyutech.jp*

When wood materials are used for a mechanical structure, the fatigue strength should be estimated due to repeated loads they receive. This paper reveals that the methods to calculate allowable stresses along with American Society for Testing and Materials (ASTM) and Architectural Institute of Japan (AIJ) can take the strength reduction due to fatigue into account because the ASTM/AIJ allowable stresses against the static strength closely resemble the fatigue limit against the wood static strength.

Keywords: Engineered wood; timber; strength; stress; mechanical design; ratio of fatigue limit.

1. Introduction

Wood materials such as lumbers and plywood are often used for mechanical devices including equipment for the elderly/disabled. Shapes of members for machines are commonly designed based on materials' fatigue strength, but for construction on creep strength. Thus, they differ in strength design. Since there is no clear standard for ratio of fatigue limit to static strength due to few references found for wood fatigue strength, the durability under a repeated load condition is difficult to be evaluated. This paper provides a consideration for the ratio of fatigue limit to wood static strength comparing the method to calculate allowable stress for American Society for Testing and Materials (ASTM) with that for Architectural Institute of Japan (AIJ).

[†]Corresponding author.
*To cite this article, please refer to its earlier version published in the *International Journal of Modern Physics B*, Volume 35, 2140026 (2021), DOI: 10.1142/S0217979221400269.

Table 1. Standard strength of wood.

		Bending		Tension parallel to grain		Compression parallel to grain	
		ASTM	AIJ	ASTM	AIJ	ASTM	AIJ
(A) Clear wood strength of White spruce with moisture content of 19%	[MPa]	34.4	—	34.4	—	16.2	—
(B) Ratio of dry		1.89	—	2.20	—	2.06	—
(C) Green clear wood strength of White spruce with moisture content of 12% = (A) × (B)	[MPa]	65.0	—	75.7	—	33.3	—
(D) Sugi strength with moisture content of 15%	[MPa]	—	33.6	—	20.4	—	27.0

2. ASTM Green Clear Wood Strength of White Spruce Corresponding to AIJ Sugi Strength

ASTM International, a standardization organization centered around North America, provides the standards ASTM D245[1] and ASTM D2555[2] for a practice to calculate wood allowable stress, in which allowable stresses are calculated from strength values of clear wood specimens. In Japan, the strength and allowable stress of Japanese wood are determined by Japanese Agricultural Standard with ASTM as a guide. AIJ provides details in standard for structural design of timber structures.[3]

Table 1 shows comparisons of calculating methods between the ASTM static strength of White spruce, American soft wood, and AIJ static strength of Sugi, Japanese soft wood similar to White spruce. Finishing materials often used as structural members are taken up here. Sugi finishing materials for structural grade 1 measure less than 36 mm in thickness, and 15% or less in humidity. White spruce shows a similar moisture content, 12%. It measures 38 mm in thickness and 140 mm in width.

3. Comparison of Allowable Stress Under Alternative Loading Between ASTM and AIJ

Table 2 shows the methods to calculate the reduction factor and allowable stress of static strength, using ASTM D245 for White spruce and AIJ for Sugi. The specifications and dimensions of materials are same as the previous section. ASTM has (a) adjustment factor, (a1) degradation factor,

Table 2. Ratio of fatigue limit to static strength for ASTM and AIJ.

	Bending		Tension parallel to grain		Compression parallel to grain	
	ASTM	AIJ	ASTM	AIJ	ASTM	AIJ
(a) Adjustment factor = (a1) × (a2)	1/2.1	(1/2.8)	1/2.1	(1/2.8)	1/1.9	(1/2.8)
(a1) Degradation factor prescribed in ASTM and AIJ	0.60	0.55	0.60	0.55	0.60	0.55
(a2) Safety factor prescribed in AIJ	(1/1.26)	2/3	(1/1.26)	2/3	(1/1.14)	2/3
(b) Strength ratio prescribed in ASTM	0.60	—	0.60 × 0.55	—	0.65	—
(c) Seasoning adjustment prescribed in ASTM	1.35	—	1.35	—	1.75	—
(d) Environmental coefficient prescribed in AIJ	—	1.0	—	1.0	—	1.0
(e) Special factor for 140 mm width prescribed in ASTM	0.89	—	—	—	—	—
(f) Product of coefficients (equivalent to ratio of fatigue limit to static strength) = (a) × (b) × (c) × (d) × (e)	0.34	0.36	0.21	0.36	0.60	0.36
(g) Allowable stress (equivalent to fatigue limit) [MPa]	22.1	12.1	15.9	7.3	20.0	9.7

(b) strength ratio, (c) seasoning adjustment, and (e) special factor. AIJ has (a1) degradation factor, (a2) safety factor, (d) environmental coefficient. Factors which are not prescribed in the standards but can be derived with calculation are shown within parentheses in Table 2.

The adjustment factor takes strength reduction due to load duration and factor of safety into account. The factor applied to clear wood properties of softwood is 1/2.1. The degradation factor is derived from Fig. 1(a) which is specified in ASTM D245. The strength with various load durations is shown

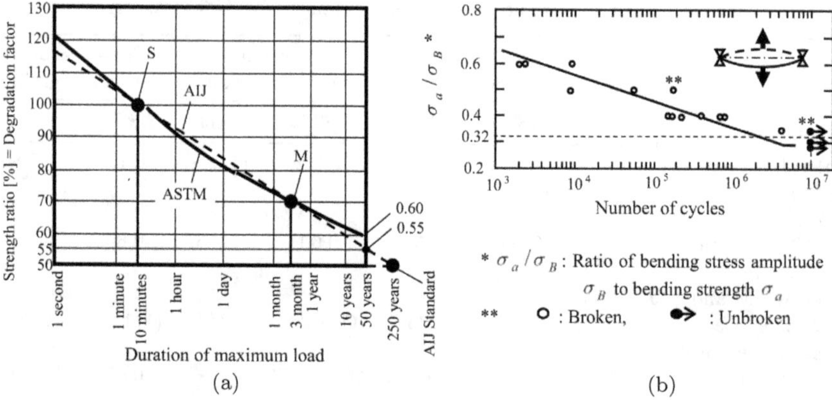

Fig. 1.   (a) Relation of strength to duration of load by ASTEM and AIJ and (b) *S–N* curve in repeated bending fatigue of Japanese Sugi wood.

referred to the static strength with 10 min as 100%. The horizontal axis is a logarithmic axis of load duration. The solid line shows ASTM values. From the viewpoint of considering wood fatigue life, the value 0.6 at the maximum load duration of 50 years in Fig. 1(a) is used for the degradation factor. The safety factor is calculated as follows. Since the adjustment factor and degradation factor are 1/2.1 and 0.6, respectively, the safety factor is $1/(2.1 \times 0.6) = 1/1.26$.

Meanwhile, AIJ specifies the degradation factor and safety factor instead of the adjustment factor. The broken line in Fig. 1(a) shows the degradation factor for AIJ. The line connecting point S at 10 min of load duration and point M at 3 months gives the point at 250 years, which AIJ sets as a standard strength. The value 0.6 at 50 years of load duration is used for the degradation factor. AIJ usually uses 2/3 for the safety factor which estimates the crushing strength as low as the elastic limit.

The strength ratio is a reduction factor that takes defects such as knots and cracks on full-sized materials into account. Tension parallel to grain is considered to be 0.55 times the reduction factor for bending because it is more sensitive to knots. The value 0.60 is used for bending strength ratio assuming that the board has a 60-mm knot in the center.

The seasoning adjustment is a factor to take moisture content of materials into account. The strength increases with moisture content decreased. When moisture content is 15% or less, the seasoning adjustment for bending and tension parallel to grain is 1.35, and that for compression parallel to grain is 1.75.

The environmental coefficient is an AIJ factor that takes degradation of wood materials including adhesive such as plywood due to UV or wetting into account. The value for the materials without adhesive is set to 1.0.

Since the strength is reduced with the size increased, the special factor is provided by ASTM D245 to reduce the strength of materials wider than 2 inches, which is applied only to the bending strength.

The product of above-mentioned coefficients, (a1), (a2), (b), (c), (d), and (e) is shown in (f) product of coefficients in Table 2. Multiplying it by the wood strength in Table 1 gives (g) allowable stress of wood materials for ASTM/AIJ in Table 2. The ASTM/AIJ values are 0.21 to 0.60 times and 0.36 times the static strength, respectively.

4. Comparison between Allowable Stress and Fatigue Limit

The product of coefficients in Table 2 corresponds to the endurance ratio, and the allowable stress in Table 2 corresponds to the fatigue limit. Figure 1(b) shows the S–N curve obtained for bending fatigue test by Imayama and Matsumoto.[4] The specimen is from a matured material of 40-year-old Sugi trunk log whose moisture content is 12% to 15%. The longitudinal direction of the specimen is the grain direction. By applying the force at the middle of the specimen, the reversed bending stress was repeated. The loading frequency was 40 cycles/s. The vertical axis in Fig. 1(b) is the ratio σ_a/σ_B, which is the bending stress amplitude σ_a against the bending strength σ_B. Based on this figure, $\sigma_{wb}/\sigma_B = 0.32$ is obtained. The ASTM/AIJ products of coefficients in Table 2 do not consider fatigue due to load fluctuation. However, ASTM allowable stress is 0.21–0.60 times the static strength, and AIJ allowable stress is 0.36 times the static strength. Since Fig. 1(b) indicates that the fatigue limit is 0.32 times the static strength, it can be considered that the ASTM/AIJ allowable stresses take fatigue into account.

5. Conclusions

In this paper, how to estimate the fatigue strength of wood materials is described in accordance with ASTM and AIJ. It was confirmed that the reduction factor corresponds to the fatigue endurance ratio, and the allowable stress corresponds to the fatigue limit. Since Japanese database does not provide American wood material data, the ASTM/AIJ methods are necessary for obtaining the fatigue limit from the ratio of fatigue limit to static strength. To estimate the tensile fatigue strength, ASTM is recommended for safety, and to estimate the compressive fatigue strength, AIJ is recommended for safety.

Acknowledgments

The authors wish to express their thanks to the members of the research group, Dr. Yoshikazu Sano, and Dr. Yasushi Takase for their kind support and advice for this study.

References

1. ASTM D245-06 (2019), Standard Practice for Establishing Structural Grades and Related Allowable Properties for Visually Graded Lumber, ASTM International, West Conshohocken, PA, 2019, www.astm.org.
2. ASTM D2555-17a, Standard Practice for Establishing Clear Wood Strength Values, ASTM International, West Conshohocken, PA, 2017, www.astm.org.
3. *Standard for Structural Design of Timber Structures* (Architectural Institute of Japan, Maruzen, Tokyo, 2006).
4. N. Imayama and T. Matsumoto, *J. Japan Wood Res. Soc.* **16**, 7 (1970).

Study of the micro/nano-texture design to improve the friction properties of DLC thin films*

Young Woo Kwon[†] and Mun Ki Bae[‡]

*Department of Nano fusion Technology, Pusan National University,
Busan 46241, Republic of Korea*
[†] *handmade4522@gmail.com*
[‡] *wck9230@naver.com*

Ri-Ichi Murakami

*School of Mechanical Engineering, Chengdu University,
Chengdu, Sichuan, P. R. China*
anewmoon816@gmail.com

Tae Hwan Jang[§] and Tae Gyu Kim[¶]

*Department of Nanomechatronics Engineering,
Pusan National University, Busan 46241, Republic of Korea*
[§] *taehwan110@hanmail.net*
[¶] *tgkim@pusan.ac.kr*

In this study, a DLC pattern was fabricated through a photolithography process that constitutes a part of the semiconductor process, to investigate the frictional wear characteristics. The photolithography was used to produce negative patterns with a pattern width of 10 μm or 20 μm and a pattern depth of 500 nm on the DLC surface. The change in the coefficient of friction of the surface was investigated through a ball-on-disk tribology test on the fabricated micro/nano-sized DLC pattern. The DLC pattern fabricated by the photolithography process showed a superior coefficient of friction to that of the general DLC sample. These results show that the decrease in the surface friction coefficient of the patterned DLC thin film is due to the reduction in the surface contact area owing to the modification of the micro/nano-texture of the surface as well as the low friction characteristics of the DLC.

Keywords: Photolithography; lift-off; ball-on-disk; DLC; wear; pattern.

[¶]Corresponding author.
*To cite this article, please refer to its earlier version published in the *International Journal of Modern Physics B*, Volume 35, 2140027 (2021), DOI: 10.1142/S0217979221400270.

1. Introduction

Recently, there has been much interest in the study of the tribology characteristics of the sliding parts of precision mechanical devices, and investigations have been focused on their lubrication characteristics, frictional resistance and wear phenomena under extreme environmental conditions.[1,2] As it is difficult to overcome the limitations of the intrinsic physical properties of most materials, the defects are compensated through physical and chemical coating processes, such as thin film deposition on the surface. Various techniques of surface control are also being studied to improve the tribology characteristics of materials.

In this study, a DLC thin film was deposited by RF-CVD on a cemented carbide (WC) material that is widely used in industrial fields to achieve high hardness and low friction characteristics. In addition, micro/nano-sized patterns were fabricated on the surface of the DLC thin film using a photolithography process.[3,4] Further, a ball-on-disk-type abrasion test was carried out under standby conditions. The purpose of this study was to improve the friction coefficient and wear characteristics according to the size and shape of the micro/nano-patterns.

2. Experimental Procedures

In this study, the cemented carbide was used as the substrate material. Test pieces were fabricated with lateral dimensions of 16 mm × 16 mm and a thickness of 5 mm. The surface of the test piece was precisely polished, and a 500-nm-thick DLC thin film was deposited. Then, a photolithography process and lift-off process were used to produce a hole-like micro/nano-sized pattern as shown Fig. 1. In the DLC deposition process, the

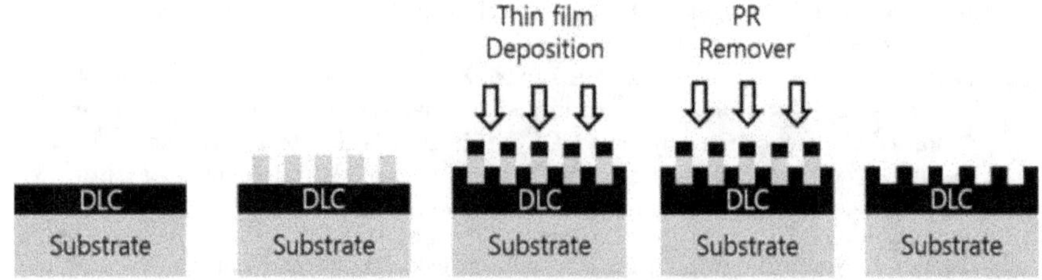

Fig. 1.   (Color online) Schematic of the photoresist lift-off process.

specimen was first mounted into the holder of the vacuum chamber, and the chamber was evacuated to an initial vacuum of 5×10^{-5} Torr. Then, plasma etching was performed for 30 min with Ar (55 sccm) at 400-W power under 4×10^{-2}-Torr pressure for 20 min. After the Ar plasma etching, 75 sccm of methane (CH_4) and 15 sccm of Ar were injected, and the vacuum pressure was maintained at 5×10^{-2} Torr. Then, the DLC thin film was deposited at an RF power of 300 W for 60 min. Afterwards, a positive photoresist (PR, Gxr 601) was spin-casted at 1000 rpm for 30 s to an optimized thickness (2 μm) on the DLC-coated (1 μm) WC substrate. In the next process, the sample was baked at 110°C for 2 min, and UV light ($\lambda = 365$ nm, I-line mask aligner) was uniformly irradiated on the entire substrate completely in contact with a Cr photomask (pattern diameter: 10 μm or 20 μm; capacity: 30 mJ $\cdot$ cm^{-2}). The development process was performed for 2 min (AZ 300MIF), and an isolated circular columnar dot-shaped pattern with the same sharp contrast as the designed photomask was formed. Then, the DLC thin film was deposited for 30 min by the DLC deposition process under the same conditions, as previously described, to ensure that the thickness of the DLC thin film was 500 nm, and the hole-type DLC pattern (diameter: 10 μm, pitch: 30 μm; or diameter: 20 μm, pitch: 60 μm) was obtained by the lift-off process.

A ball-on-disk wear test was conducted under standby conditions using the following test conditions: a load of 1 N, wear distance of 1000 m, wear speed of 120 rpm and wear track radius of 7 mm. An Al_2O_3 ball of ø3 was used as the relative material. The diameter and pitch of the hole patterns on the DLC films were confirmed by optical microscopy, as shown in Fig. 2.

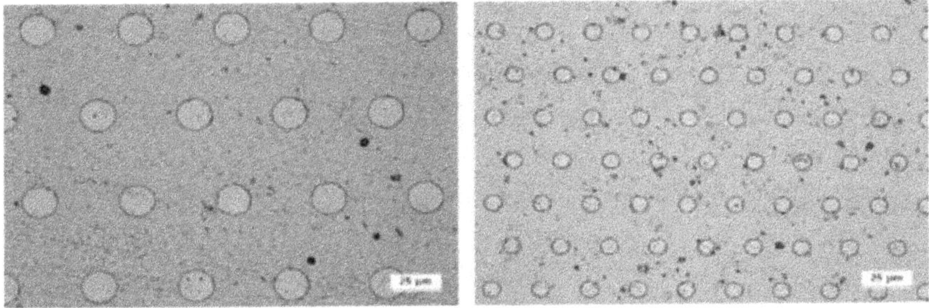

Fig. 2.   (Color online) Optical microscopy images of the DLC surfaces with hole patterns by photolithography.

3. Results and Discussion

In the patterning process, the adhesion between the substrate material and DLC thin film is very important. A scratch test was performed to measure the adhesion of the DLC pattern test piece. Lc1 from which the initial peeling of the DLC thin film started was measured to be 8.21 N, and Lc2 from which the DLC thin film was completely peeled was found to be 36.1 N. From this result, it was judged that the adhesion of the patterning DLC specimen was very good. The hardness of the cemented WC substrate was measured to be HV 2530, whereas the hardness of the DLC thin film deposited on the cemented WC substrate was measured to be higher at HV 3212, which indicated that the surface hardness was improved.

Figure 3 shows the experimental results for the behavior of the coefficient of friction of patterned specimens subjected to the ball-on-disk tribology test. The average coefficient of friction of the nonpatterned DLC thin film was measured to be 0.125. The average coefficient of friction of the specimen patterned with a pattern diameter of 20 μm was 0.102, and the coefficient of friction of the specimen patterned with a pattern diameter of 10 μm was further reduced to 0.068. Therefore, in the ball-on-disk wear test, it was found that the narrower the pitch of the pattern and the smaller the diameter of the pattern, the lower the average coefficient of friction. This result shows that the friction coefficient decreases as the contact

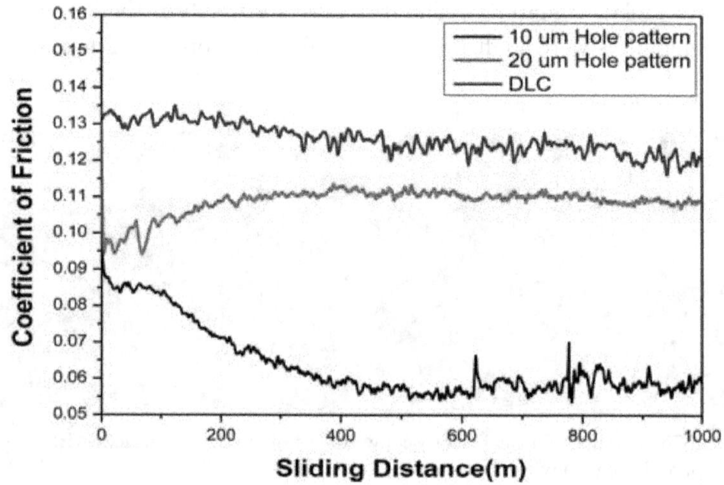

Fig. 3.   (Color online) Average friction coefficients of different DLC specimens.

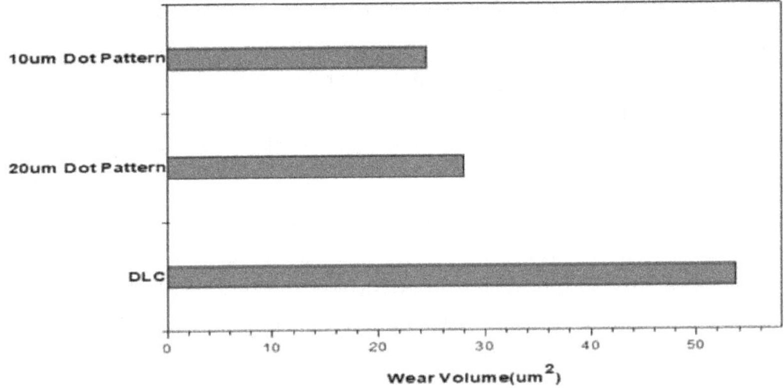

Fig. 4. (Color online) Measured wear volumes after the friction test for patterned specimens.

area where the counterpart Al_2O_3 ball directly rubs against the DLC thin film during the wear test decreases.

The friction coefficient of the patterned sample with a pattern diameter of 10 μm was found to be 12.5% lower than that of the sample with a larger pattern diameter of 20 μm, and 54.4% lower than that of the general DLC thin film. The coefficient of friction of the sample with holes of 20-μm diameter was 47.8% lower than that of the general DLC thin film. Thus, it was found that the smaller the spacing and diameter of the pattern are, the better will be the coefficient of friction characteristics.

Further, the wear area was analyzed using a 2D profile-meter. The 2D profile-meter is a method for calculating the wear area by precisely measuring the wear width and wear depth of a surface. Figure 4 compares the wear areas of the three types of DLC films. The wear area of the nonpatterned DLC sample was found to be 53.7 μm^2, while that of the sample with 20-μm pattern diameter was 28 μm^2, and that of the sample with 10-μm pattern diameter was 24.5 μm^2. These results confirmed that the lower the coefficient of friction, the smaller the amount of wear.

4. Conclusion

Our conclusions are as follows:

(1) A DLC thin film was deposited on a WC surface, and a hole-type micro/nano-pattern was fabricated using the photolithography process

and lift-off process, and the friction characteristics of the patterned films were investigated.

(2) A ball-on-disk wear test indicated that the friction coefficients of the patterned samples with holes of 10-μm diameter (0.068) and 20-μm diameter (0.102) decreased by 45.6% and 18.4%, respectively, compared to that of the nonpatterned DLC thin film (0.125). It was found that when the spacing of the pattern and the size of the diameter decreased, the friction characteristics were excellent.

(3) The wear area was analyzed by 2D profile-meter, and it was confirmed that a lower friction coefficient leads to a smaller wear amount.

References

1. K. G. Foote and D. N. MacLennan, *J. Acoust. Soc. Am.* **75**, 612 (1984).
2. F. Cardarelli, Ceramics, refractories, and glasses, in *Materials Handbook: A Concise Desktop Reference* (Springer Science & Business Media, 2008), pp. 593–689.
3. T. Roch, V. Weihnacht, H.-J. Scheibe, A. Roch and A. F. Lasagni, *Diam. Relat. Mater.* **33**, 20 (2013).
4. M. K. Bae, H. G. Lim, Y. M. Park, S. S. Kim and T. G. Kim, *Mod. Phys. Lett. B* **34**, 2040044 (2020).

Accurate stress intensity factors for kinked interface crack in bonded dissimilar half-plane[*]

Kazuhiro Oda[†]

Division of Mechanical Engineering,
Faculty of Science and Technology, Oita University 700 Dannoharu,
Oita-shi 870-1192, Japan
oda-kazuhiro@oita-u.ac.jp

Nao-Aki Noda

Department of Mechanical Engineering,
School of Engineering, Kyushu Institute of Technology Kitakyushu-shi,
Fukuoka 804-8550, Japan
noda.naoaki844@mail.kyutech.jp

In this study, the stress intensity factor (SIF) of an interface kinked crack is analyzed by the singular integral equation of the body force method. The problem can be expressed by distributing the body force doublets of the tension and shear types along all the boundaries of the kinked and interface crack parts. The SIFs can be obtained directly from the densities of the body force doublets at the crack tips. Although the problem has already been calculated using the crack connection model, the accuracy of the analysis has not been clarified. From the analysis results in this study, it can be seen that the SIFs calculated by the crack connection model have a nonnegligible error, and the present method gives more accurate results. The advantage of the present method is that the SIFs of the kinked and the interface crack tips can be obtained at the same time with high accuracy.

Keywords: Stress intensity factor; kinked interface crack; singular integral equation.

[†]Corresponding author.
[*]To cite this article, please refer to its earlier version published in the *International Journal of Modern Physics B*, Volume 35, 2140030 (2021), DOI: 10.1142/S0217979221400300.

1. Introduction

In recent years, composites and adhesively bonded materials have been used in a wide range of engineering applications. The problems associated with interfacial cracks are of great interest because defects and cracks along the interface can reduce the strength of the structure.[1-6] However, there are only few detailed studies on the stress intensity factors (SIFs) of kinked interface cracks,[3,5] which are of practical importance, and the accuracy of the analysis has not been clarified.

We have previously reported that the SIFs for kinked cracks in a homogeneous plane[7] and collinear interface cracks in a bonded dissimilar plane[6] can be accurately calculated by the numerical solution of singular integral equation of the body force method (BFM). In this paper, the method is applied to the kinked interface crack problem. By comparing the present results with those of other studies, the accuracy of the proposed method is demonstrated. The effects of kinked angle and material combination on the SIF are also discussed.

2. Numerical Solution of Singular Integral Equation

In this study, the problem of a kinked interface crack in dissimilar materials is treated as shown in Fig. 1. The elastic constants are given as shear moduli and Poisson's ratios for the upper (material 1) and the lower (material 2)

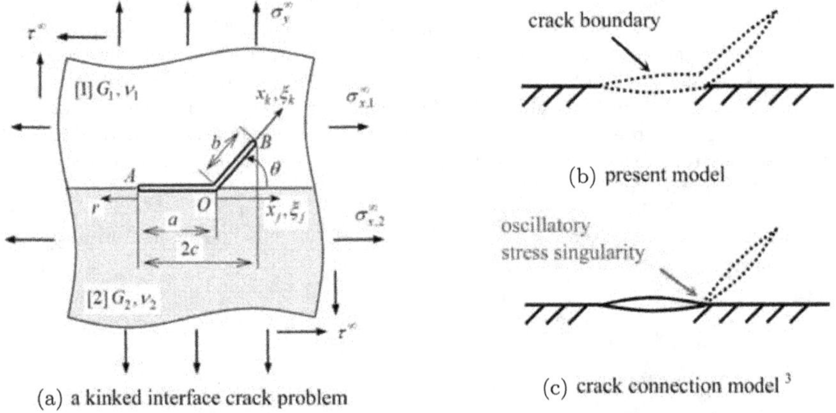

Fig. 1.   (Color online) Treated problem and crack boundary distributed force doublets: (a) a kinked interface crack problem, (b) present model and (c) crack connection model in Ref. 3 to satisfy the boundary condition of crack surface.

half-planes, i.e., (G_1, ν_1) and (G_2, ν_2). From the perfect bonding condition of interface, the remote stresses have the following relation[3]:

$$\sigma_{x,2}^{\infty} = \frac{G_2}{G_1}(1 + \kappa_1)\sigma_{x,1}^{\infty} + \left\{3 - \kappa_2 - \frac{G_2}{G_1}(3 - \kappa_1)\right\}\sigma_y^{\infty}. \tag{1}$$

Here, $\kappa_m = (3 - \nu_m)/(1 + \nu_m)$ for plane stress and $\kappa_m = 3 - 4\nu_m$ for plane strain $(m = 1, 2)$.

The problem can be formulated in terms of singular integral equations by means of the BFM.[6,7] In the interface crack problem, the stress fields induced by two kinds of standard set of force doublets, tension type and shear type, in dissimilar materials without the crack are used as the fundamental solution.[6,7] The singular integral equations can be obtained by the force doublets along the imaginary boundaries of the crack as shown in Fig. 1(b). The SIFs can be directly evaluated from the densities of body force doublets distributed on the imaginary interface crack part and kinked crack part, respectively.[6,7]

This problem has been already analyzed by using the crack connection model as shown in Fig. 1(c).[3] In this model, the boundary condition of the interface crack part is completely satisfied because the fundamental solution of bonded half-planes with an interface crack is used. However, the accuracy of the obtained SIFs has not yet been clarified. As shown in the following, the present method gives more accurate results than using the crack connection model.

3. Numerical Results and Discussion

The problem of the kinked interface crack is analyzed when the relative kinked crack length b/a and the material combination G_2/G_1 are changed systematically. In this analysis, Poisson's ratios $\nu_1 = \nu_2 = 0.3$ and the plane stress condition is assumed. The normalized SIFs, $F_{1,A}$, $F_{2,A}$, $F_{I,B}$ and $F_{II,B}$, are defined by the following expressions:

$$K_{1,A} + iK_{2,A} = \{F_{1,A} + iF_{2,A}\}\sigma\sqrt{\pi a}(1 + 2i\varepsilon), \tag{2}$$

$$K_{I,B} = F_{I,B}\sigma\sqrt{\pi b}, \quad K_{II,B} = F_{II,B}\sigma\sqrt{\pi b}. \tag{3}$$

Here, $K_{1,A}$ and $K_{2,A}$ are the SIFs of the interface crack defined by[2]

$$\sigma_y + i\tau_{xy} = \frac{K_{1,A} + iK_{2,A}}{\sqrt{2\pi r}}\left(\frac{r}{2a}\right)^{i\varepsilon}, \quad \varepsilon = \frac{1}{2\pi}\ln\left(\frac{G_2\kappa_1 + G_1}{G_1\kappa_2 + G_2}\right) \tag{4}$$

where r is the distance from the interface crack tip shown in Fig. 1(a).

Table 1 shows the comparison of numerical results between the present model and the crack connection model illustrated in Fig. 1. From Table 1, when $G_2/G_1 = 1$, both results are in good agreement with each other. However, for $G_2/G_1 = 0.25$ and 4, there is a maximum difference of about 10% between both the results. The values in parentheses are the results analyzed by the finite element method (FEM) for $b/a = 1$. The FEM values are close to the present results. Therefore, the numerical results obtained in this study are more accurate than the values calculated by the crack connection model.

Figures 2 and 3 show the normalized SIFs of the interface crack tip A and the kinked crack tip B under three types of loading conditions, respectively. The analysis is performed by changing the kinked angle θ and G_2/G_1.

Table 1.   Comparison of normalized SIF, $F_{\text{I},B}$ ($\sigma_y^\infty = 1, \theta = 45°$).

| G_2/G_1 | 0.25 | | 1 | | 4 | |
| | Present model [Fig. 1(b)] | Ref. 3 [Fig. 1(c)] | Present model [Fig. 1(b)] | Ref. 3 [Fig. 1(c)] | Present model [Fig. 1(b)] | Ref. 3 [Fig. 1(c)] |
b/a						
0.1	0.743	0.703	0.655	0.655	0.634	0.705
0.5	0.800	0.770	0.663	0.663	0.620	0.668
1	0.902	0.877	0.743	0.744	0.690	0.732
1 (FEM)	(0.899)		(0.743)		(0.687)	
1.5	1	0.977	0.824	0.824	0760	0.799

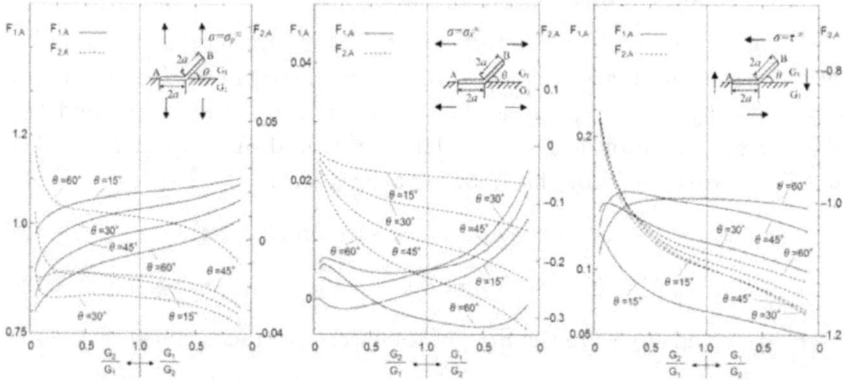

Fig. 2.   Normalized SIFs $F_{1,A}$ and $F_{2,A}$ for kinked interface crack at the interface crack tip A under three different loads: (a) tension in y-direction, (b) tension in x-direction and (c) shear when $b/a = 1$, $\nu_1 = \nu_2 = 0.3$ and under plane stress.

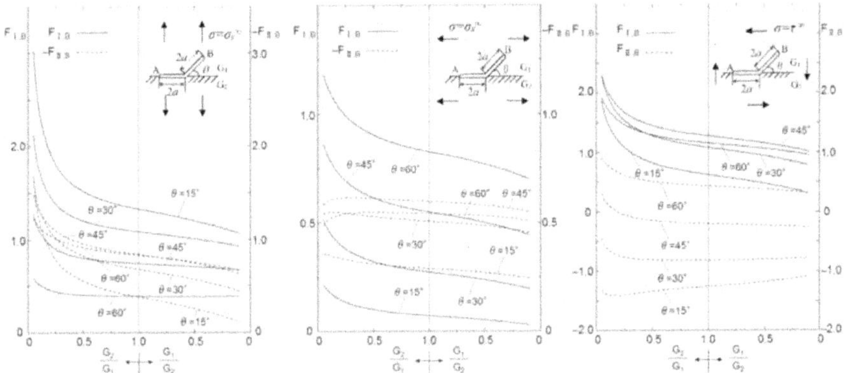

Fig. 3. Normalized SIFs $F_{I,B}$ and $F_{II,B}$ for kinked interface crack at the kinked crack tip B under three different loads: (a) tension in y-direction, (b) tension in x-direction and (c) shear when $b/a = 1$, $\nu_1 = \nu_2 = 0.3$ and under plane stress.

As shown in Fig. 2, $F_{1,A}$ and $F_{2,A}$ of interface crack tip A under tension in the y-direction are $0.75 < (F_{1,A}, F_{2,A}) < 1.2$ even when G_2/G_1 changes significantly. From Fig. 3, $F_{I,B}$ and $F_{II,B}$ for kinked crack tip B are less affected by the stiffness ratio when $G_2/G_1 \geq 0.5$, regardless of the loading conditions.

4. Conclusion

In this paper, the SIFs for kinked interface crack were calculated accurately by using the singular integral equations of the BFM. By comparing the present results with those of other methods, it was found that the present method gives more accurate results for the kinked interface crack. The normalized SIFs at the kinked crack tip are less affected by the stiffness ratio when $G_2/G_1 \geq 0.5$, regardless of the loading conditions.

References

1. F. Erdogan, *J. Appl. Mech.* **30**, 232 (1963).
2. R. Yuuki and S. B. Cho, *Eng. Fract. Mech.* **34**, 179 (1989).
3. M. Isida and H. Noguchi, *Eng. Fract. Mech.* **46**, 365 (1993).
4. H. Nisitani, A. Saimoto and H. Noguchi, *Trans. JSME* **59A**, 68 (1993) (in Japanese).
5. L.-G. Zhao and Y.-H. Chen, *Int. J. Fract.* **76**, 279 (1996).
6. N.-A. Noda and K. Oda, *Int. J. Fract.* **84**, 117 (1997).
7. N.-A. Noda and K. Oda, *Int. J. Fract.* **58**, 285 (1992).

Quantitative analysis of surfaces of cross-section in butt welding of aluminum alloy 5083 using laser–arc hybrid welding*

Ji-Sung Kim[†]

*Korea Marine Equipment Research Institute, 24-20,
Noksansandan 335-ro, Gangseo-Gu Busan 46754, South Korea
jskim@komeri.re.kr*

Jong-Do Kim

*Division of Marine Engineering, Korea Maritime
and Ocean University, 727 Taejong-ro, Yeongdo-Gu,
Busan 49112, South Korea
jdkim@kmou.ac.kr*

The alloy aluminum of 5000 series will have relatively high strength through solid solution strengthening of Mg. However, when laser welding the 5000 series aluminum alloys, the Mg is selectively evaporated by welding heat due to its low melting and vaporization points, resulting in a reduction in the strength of welds. Therefore, laser welding application is difficult because of such a reduction in strength[1,2] [M. Peel, A. Steuwer, M. Preuss and P. J. Withers, *Acta Mater.* **51**, 4791 (2003); A. Haboudou, P. Peyre, A. B. Vannes and G. Peix, *Mater. Sci. Eng. A* **363**, 40 (2003)]. In this paper, welding experiments were carried out using laser and laser–arc hybrids on aluminum alloy 5083 with 8-mm thickness. Tensile testing, scanning electron microscopy (SEM) analysis and electron probe microanalyzer (EPMA) analysis were carried out on specimens on which the laser and laser–arc hybrid weldings were performed. According to the tensile test results, the tensile strength during laser–arc hybrid welding was greater than 85% of the parent metal. In addition, EPMA analysis showed that the strength of the laser–arc hybrid welds was maintained by supplementing the optional evaporating Mg element, since they use the filler wire.

Keywords: Aluminum alloy; A5083; laser welding; laser–arc hybrid welding.

[†]Corresponding author.

*To cite this article, please refer to its earlier version published in the *International Journal of Modern Physics B*, Volume 35, 2140031 (2021), DOI: 10.1142/S0217979221400312.

1. Introduction

The greatest advantages of aluminum are its light weight and excellent mechanical properties. Furthermore, aluminum has durability based on corrosion resistance and its excellent formability and machinability facilitate the production of products. Therefore, the utilization of aluminum is increasing in the fields of bicycles, high-speed rail vehicles, aircraft fuselage and ships. The aluminum alloy 5083 of the 5000 series is an Al–Mg alloy that maintains mechanical strength through the solid solution strengthening effect of Mg. Thus, it has the highest strength among the nonthermal treatment alloys. Therefore, aluminum alloy 5083 with high strength is applicable to almost all parts of an aluminum ship, such as side of the ship, external board of ship bottom, rip plate, separating panel, rib, engine pedestal, operational room funnel, deck and bulwark. Active researches on aluminum alloy welding are underway with the replacement of fiber-reinforced plastic (FRP) ships by aluminum ships and the increasing demand for cruise ships and yachts. Peel *et al.* reported that heat input determines the welding properties than the shape of tool in the friction stir welding of aluminum alloy 5083.[1] Haboudou *et al.* applied dual spot to suppress pores generated by the laser welding of aluminum alloys and reported that it was effective in suppressing pores.[2] Ancona *et al.* evaluated the weldability of laser welding with 3-mm-thick aluminum alloy 5083 according to the shield gas nozzle shape.[3] Chang *et al.* investigated the causes of pores formed during laser welding and found that pores decreased at higher laser power or lower welding speed.[4] In this study, arc, laser and laser–arc hybrid weldings were performed on 8-mm-thick aluminum alloy 5083 sheets and their weldabilities were compared and the effects of alloying elements on weldability were examined.

2. Materials and Method

The material used in this study is aluminum alloy 5083, and sheets with the sizes of 150 mm in length, 300 mm in width and 8 mm in thickness were used (Table 1). Laser and laser–arc hybrid weldings were performed to examine

Table 1. Chemical compositions of aluminum alloy 5083 (wt.%).

Material \ Elements (wt.%)	Si	Fe	Cu	Mn	Mg	Cr	Ni	Zn	Ti	Al
A5083-H321	0.12	0.29	0.011	0.6	4.59	0.10	0.0044	0.0093	0.023	Bal.
ER5356	0.25	0.40	0.10	0.05–0.20	4.5–5.5	0.05–0.20	—	0.10	0.06—0.20	Bal.

the effects of alloying elements on weldability. Laser welding was performed by two passes in total. Power outputs of 5.6 kW (front) and 3 kW (back) and a welding speed of 2 m/min were used here. In the case of laser–arc hybrid welding, an arc current of 120 A and the same welding conditions as for laser welding were used. For the filler metal for arc and laser–arc hybrid weldings, ER5336, which contains 4.5–5.5% of Mg, was used.

3. Results and Discussion

3.1. *Tensile test results of laser and laser–arc hybrid welds*

Tensile tests were performed for two conditions to compare the tensile properties of laser weld and laser–arc hybrid weld. Figure 1 shows the fracture shapes and the scanning electron microscopy (SEM) images of the fracture surfaces as the results of tensile test. The tensile test results showed that the tensile strength of laser weld was 217 MPa, which was approximately 70% of that of the parent metal. However, the laser–arc hybrid weld showed a tensile strength of 262 MPa, which was 85% of that of the parent metal. In general, according to classification rules, the weld must have a strength of 85% or more of the parent metal to receive "pass" judgment. Thus, the laser weld does not satisfy the pass criterion. The fracture position of the weld was examined, and the fracture of the laser weld occurred at the melting part. By contrast, in the case of laser–arc hybrid weld, the fracture occurred at the heat-affected zone. The fracture type was ductile fracture for both the welds. The SEM images of the fracture surface showed many pores for both laser and laser–arc hybrid welds. The laser weld showed

Tensile test specimen	Fracture image	SEM image	Fracture type /position
Laser welding		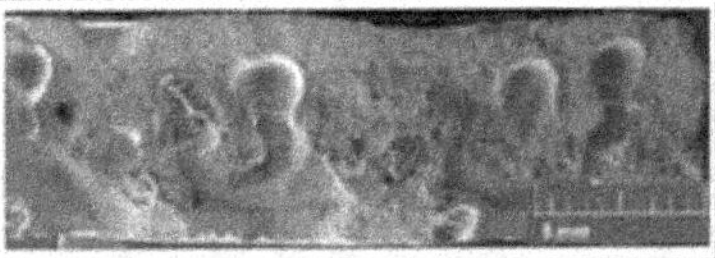	Ductile/ Weld metal zone
Laser-arc hybrid		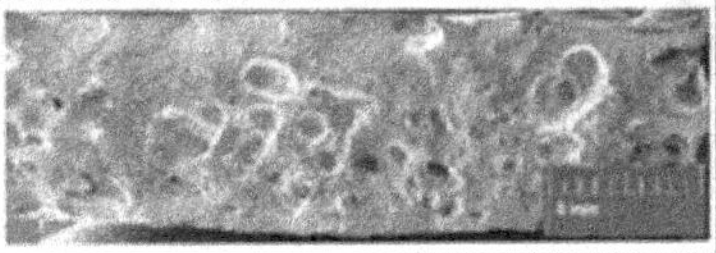	Ductile/ Heat Affected zone

Fig. 1. Results of tensile–shear tests for the specimens of laser and laser–arc hybrid weldings.

pores of large elongated shapes, whereas the laser–arc hybrid weld showed smaller circular pores.

3.2. *EPMA analysis results of laser and laser–arc hybrid welds*

In order to examine the cause of decreased strength of laser weld and the effects of alloying elements on the weldability of laser–arc hybrid welding, quantitative analysis of laser and laser–arc hybrid welds was performed using an electron probe microanalyzer (EPMA). The quantified alloying elements include Fe, Si, Ni, Mn, Mg, Cr, Al, Cu and Zn. The distributions of alloying elements of the weld were expressed by colors. The closer the color to red, the higher the content, and the closer it is to blue, the lower the content. Figures 2 and 3 show the EPMA analysis results of the laser weld and laser–arc hybrid weld, respectively. The distribution trend of Mg for laser weld indicates that the Mg content of the melting part was lower than that of the parent metal, and this trend was more distinct towards the bottom. On the other hand, the distribution trend of Mg for laser–arc hybrid weld indicates that unlike the laser weld, the Mg content difference

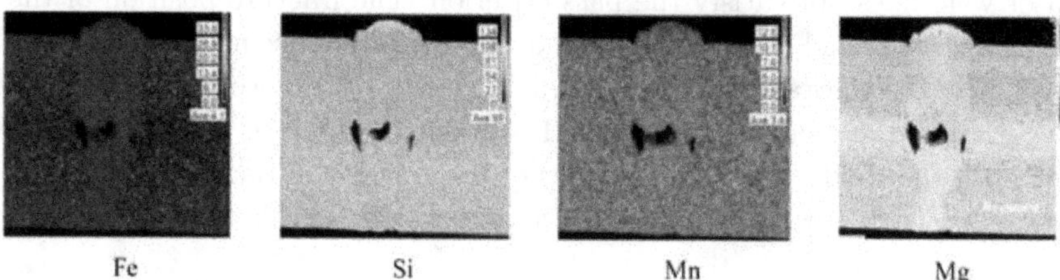

Fig. 2. (Color online) Results of EPMA analysis at the cross-section using laser welding.

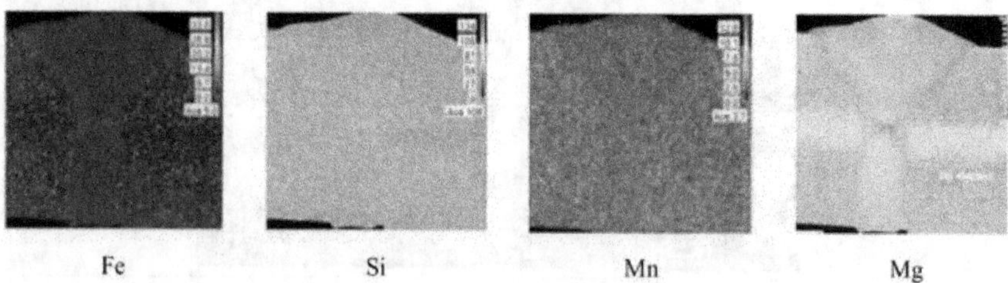

Fig. 3. (Color online) Results of EPMA analysis at the cross-section using laser–arc hybrid welding.

between the melting part and parent metal is not large. The quantification analysis using EPMA showed that laser welding of aluminum alloy 5083 decreases the Mg content of the melting part because the Mg elements are evaporated selectively. Furthermore, in the case of laser–arc hybrid welding, the reduction of Mg content of the melting part could be suppressed because the Mg elements were supplemented by the filler metal. Consequently, the strength decrease of the laser weld is considered to be caused by the selective evaporation of the Mg elements during welding. This confirms that when laser–arc hybrid welding using a filler metal is applied, the strength decrease can be suppressed because the evaporated Mg elements are supplemented.

4. Conclusion

The butt-welding weldability of 8-mm-thick aluminum alloy 5083 sheets was evaluated using laser and laser–arc hybrid weldings and the results are summarized as follows.

(1) The laser welding of aluminum alloy 5083 did not satisfy the pass criterion of the classification rules due to a decrease in the tensile strength. However, the laser–arc hybrid welding satisfied the pass criterion of the classification rules because the tensile strength was 262 MPa (85% of the parent metal).

(2) Quantification analysis using an EPMA was performed to examine the effects of alloying elements on weldability. According to the results, the laser weld showed a decrease in the Mg content in the melting part. In the case of the laser–arc hybrid weld, the Mg distribution was not much different from that of the parent metal.

(3) It was found that the strength decrease of the laser weld was due to the selective evaporation of Mg and that the strength decrease can be suppressed by applying laser–arc hybrid welding using a filler metal.

References

1. M. Peel, A. Steuwer, M. Preuss and P. J. Withers, *Acta Mater.* **51**, 4791 (2003).
2. A. Haboudou, P. Peyre, A. B. Vannes and G. Peix, *Mater. Sci. Eng. A* **363**, 40 (2003).
3. A. Ancona, T. Sibillano, L. Tricarico, R. Spina, P. M. Lugara, G. Basile and S. Schiavone, *J. Mater. Process. Technol.* **164–165**, 971 (2005).
4. B. Chang, C. Allen, J. Blackburn and P. Hilton, *Phys. Procedia* **41**, 478 (2013).

Photo-enhanced field-emission behavior of CdSSe microflowers*

Amol B. Deore[†] and Mahendra A. More[‡]

*Department of Physics, Savitribai Phule Pune University,
Pune 411007, Maharashtra, India*
[†] *deoreamol91@gmail.com*
[‡] *mam@physics.unipune.ac.in*

Bhausaheb B. Musmade

*D. Y. Patil College of Engineering, Savitribai Phule Pune University, Akurdi,
Pune 411044, Maharashtra, India*
bbmusmade@dypcoeakurdi.ac.in

Sachin D. Nerkar

*Shri Vile Parle Kelavani Mandal's Institute of Technology,
Dhule 424001, Maharashtra, India*
snerkar2401@gmail.com

Padmakar G. Chavan

*School of Physical Sciences,
Kavuyitri Bahinabai Chaudhari North Maharashtra University,
Jalgaon 425001, Maharashtra, India*
pgchavan@nmu.ac.in

Pankaj M. Koinkar[§]

*Department of Optical Science,
Graduate School of Advanced Technology and Science,
Faculty of Engineering Tokushima University, 2-1 Minamijosanjima-cho,
Tokushima 770-8506, Japan*
koinkar@tokushima-u.ac.jp

In this work, we synthesized well-grown cadmium sulfoselenide (CdSSe) microflowers on gold-coated silicon substrate using a simple and low-cost

---

[§]Corresponding author.

*To cite this article, please refer to its earlier version published in the *International Journal of Modern Physics B*, Volume 35, 2140032 (2021), DOI: 10.1142/S0217979221400324.

chemical bath deposition technique. The deposited CdSSe film was annealed in a furnace for 30 min at 250°C. Subsequently, the annealed CdSSe film was further characterized using structural, morphological and field-emission analyses. The CdSSe microflowers show photosensitive field-emission behavior. In the photosensitive field emission, the emission current rises by nearly 2–3 times of its primary preset current value of ~ 1 μA.

Keywords: Cadmium sulfoselenide; Na_2Se_3; field emission; microflowers; current stability.

1. Introduction

Chalcogenide semiconductors have received enormous attention due to their unique chemical, mechanical and electro-optical properties and they are used in a variety of applications in commercial products. A chalcogenide semiconductor belonging to II–VI semiconductor group shows better performance in terms of luminescence, photocatalysis, photoelectrochemistry, etc.[1] The ternary alloy semiconductors differ from binary semiconductors because of their band-tunable property.[2] For example, CdSSe is an eminent promising ternary semiconductor in the cadmium (Cd) chalcogenide group.

Field emission (also called cold emission) is nothing but a quantum mechanical process in which electrons are tunneled through the surface of a material under a strong electrostatic field.[3] In the last few decades, the field-emission properties of 3D micro/nanostructures as well as field emitters have been well studied.[4] The low turn-on field, high current density at a minimum voltage and constant current stability are the major parameters of the field-emission study. These parameters mainly depend on tailored morphology (shape and size) and electrical conductivity of the materials.[5]

In this study, we report the synthesis of CdSSe microflowers on gold (Au)-coated silicon substrate using chemical bath deposition (CBD) technique at room-temperature and study their photosensitive field emission under visible-light illumination. To the best of our knowledge, photosensitive field emission of CdSSe material has not been reported till date.

2. Experimental

The CdSSe film was prepared using cadmium acetate $[C(CH_3COO)_2]$, aqueous ammonia $(NH_3 \cdot H_2O)$, thiourea $[SC(NH_2)_2]$ and sodium selenosulfate (Na_2SeSO_3). All AR-grade chemicals were purchased from Sigma Aldrich and used without further purification. We add 13 mL of 0.25-M $Cd(CH_3COO)_2$ solution into a 100-mL-capacity glass beaker. The $NH_3 \cdot H_2O$ was dropwise added to the prepared solution with a constant magnetic stirring in order to make the solution clear and transparent. In addition to this, we prepare 13 mL of 0.25-M $SC(NH_2)_2$ and Na_2SeSO_3 solutions and these were gently added into the aqueous solution of $Cd(CH_3COO)_2$ with constant stirring, respectively. The pre-cleaned Au-coated Si substrate was eventually immersed at an appropriate angle into the beaker having the clear and transparent solution. The reaction takes place at room-temperature for 30 h in the absence of mechanical stirring. After completion of the reaction, the obtained substrate was removed from the precursors, rinsed with double deionized water and then dried at room-temperature. The as-synthesized film was annealed in the furnace at 250°C for 30 min.

3. Results and Discussion

3.1. *Structural analysis*

The X-ray diffraction (XRD) spectra of annealed CdSSe films were recorded over the 2θ values from 20° to 80°. The observed five diffraction peaks (as shown in Fig. 1) are indexed as (002), (101), (102), (110) and (112) planes of the hexagonal crystalline phase of CdSSe film (JCPDS No. 49-1459). All observed diffraction peaks are compatible with the published data,[6] showing the formation of material in a pure phase. The XRD patterns evidently show that all diffraction peaks have considerable broadening, suggesting the nanocrystalline nature of the sample. The broadening of the diffracted peaks arises due to the small grain size of the particles.[7]

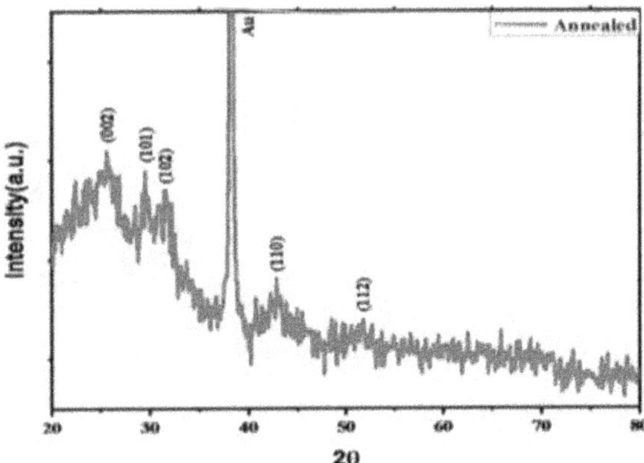

Fig. 1.   (Color online) XRD patterns of CdSSe microflowers.

3.2. *Surface morphology*

The surface micrographs of annealed CdSSe microflowers examined using scanning electron microscopy (SEM) with different magnifications are displayed in Figs. 2(a) and 2(b). The SEM images of CdSSe film show that the microflowers appear to be discrete with uniform decorated nanoflakes, which maintain hierarchical flower-like structures. CdSSe microflower edges are very "sharp", thin and transparent and the thickness of the nanoflakes is estimated to be in the order of 31–87 nm. We can observe well-aligned growth of microflowers in which the length of nanoflakes with sharp edges is advantageous for better field-emission properties.[8]

3.3. *Photo-assisted field emission*

CdSSe shows photoconductive behavior due to the tuning of bandgap in the visible region.[9] Figure 3(a) depicts the photo-enhanced FE current versus time characteristic under light illumination on the cathode surface (i.e., ITO coated glass). The intensity of the halogen lamp on the emitter surface is found to be $\sim$80 W/m^2 and is calibrated by using a photocell at a distance of $\sim$10 cm between the cathode and the lamp. At short time intervals of light "ON–OFF" condition (ON = 2 min and OFF = 3 min), rapid increment and decrement in current value can be observed [as shown in Fig. 3(a)]. The maximum increase in the current level is observed to be 2–3 times of the preset current value of $\sim$1 μA. After light illumination, the current value increases due to the generation of electron–hole pairs. The effect of photo-enhanced field-emission current of CdSSe microflowers is seen to be reproducible. The plot of J versus E, i.e., the emission current density of

Fig. 2.   SEM images of CdSSe microflowers at (a) low magnification and (b) high magnification.

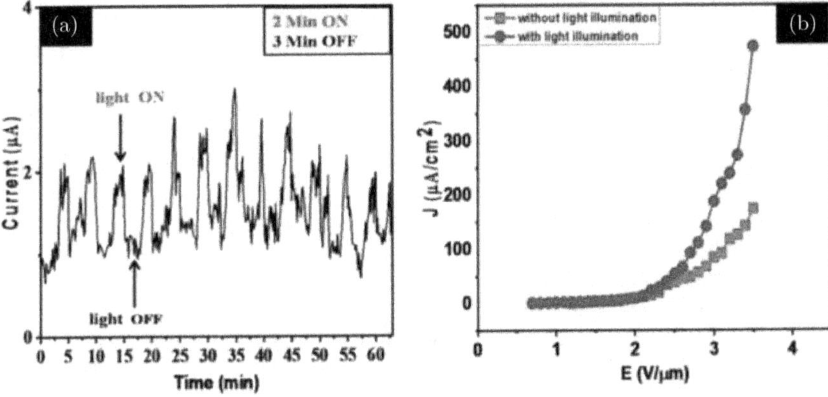

Fig. 3. (Color online) (a) Plot of current versus time (I–t) for the photoswitching behavior of CdSSe microflowers, at the preset current value of $\sim$1 μA. (b) J–E plot of CdSSe microflowers with and without light illumination.

Table 1. Turn-on fields and maximum current densities of CdSSe microflowers with and without light illumination.

Field emitter	Mode of operation	Turn-on field (preset current density value at $\sim$1 μA $\cdot$ cm^{-2})	Maximum current density
CdSSe microflowers	With light illumination	1 V $\cdot$ μm^{-1}	472.4 μA $\cdot$ cm^{-2} at a field of $\sim$3.5 V $\cdot$ μm^{-1}
	Without light illumination	1.4 V $\cdot$ μm^{-1}	173.6 μA $\cdot$ cm^{-2} at a field of $\sim$3.5 V $\cdot$ μm^{-1}

the emitter with and without the illumination of light, is shown in Fig. 3(b). From Table 1, we can see that there is an enhancement in turn-on field and current density after light illumination.

4. Conclusion

In summary, hierarchical-like CdSSe microflowers have been auspiciously synthesized by chemical bath deposition at room-temperature. When exposed to light illumination, the CdSSe microflowers reduced the turn-on field and enhanced the field-emission current density due to photoconductivity of CdSSe. The promising result of photosensitive field emission of microflowers makes it a potential candidate for application in vacuum micro/nanoelectronic devices.

References

1. P. Mandal, R. S. Srinivasa, S. S. Talwar and S. S. Major, *Appl. Surf. Sci.* **254**, 5028 (2008).
2. T. P. Tang, W. L. Wang and S. F. Wang, *J. Alloys Compd.* **488**, 250 (2009).
3. E. L. Murphy and R. H. Good, *Phys. Rev.* **102**, 1464 (1956).
4. R. B. Kamble, N. Tanty, A. Patra and V. Prasad, *Appl. Phys. Lett.* **109**, 083102 (2016).
5. S. Dhongade *et al.*, *ACS Appl. Nano Mater.* **3**, 9749 (2020).
6. K. V. Khot *et al.*, *New J. Chem.* **38**, 5964 (2014).
7. K. V. Khot *et al.*, *RSC Adv.* **5**, 40283 (2015).
8. V. S. Bagal *et al.*, *RSC Adv.* **6**, 41261 (2016).
9. J. Lu, H. Liu, S. Lim, S. Tang, C. Sow and X. Zhang, *J. Phys. Chem. C* **117**, 12379 (2013).

Study on structural design and analysis of 18 ft CFRP leisure boat*

Sung-Youl Bae

*Fibrous Ceramics & Aerospace Materials Center,
Korea Institute of Ceramic Engineering and Technology, Korea*
bsy@kicet.re.kr

Yun-Hae Kim[†]

*Department of Ocean Advanced Materials Convergence Engineering,
Korea Maritime and Ocean University, Korea*
yunheak@kmou.ac.kr

The purpose of this study is to develop a lightweight design model for an 18ft leisure boat. The existing leisure boat is manufactured using glass fiber-reinforced plastics (GFRP) material and the hand lay-up process. Carbon fiber-reinforced plastics (CFRP) was applied to the new design to reduce the boat's weight, while an automated tape laying machine was applied to the lightweight boat's manufacturing process to increase boat manufacturing productivity. The newly designed CFRP model is 25% lighter than the existing GFRP model. It was confirmed that the newly designed lightweight hull has sufficient structural integrity compared to the existing hull through the structural integrity evaluation by the FEA.

Keywords: Leisure boat; CFRP; automated tape laying; lightweight; composites.

1. Introduction

As the International Maritime Organization (IMO) has announced air pollutant emission regulations since 2016, it is essential to reduce ships and equipment parts' weight. Currently, glass fiber-reinforced plastics (GFRP) are mainly applied as lightweight materials for medium and small-sized

---

[†]Corresponding author.
*To cite this article, please refer to its earlier version published in the *International Journal of Modern Physics B*, Volume 35, 2140036 (2021), DOI: 10.1142/S0217979221400361.

ships, whereas carbon fiber-reinforced plastics (CFRP), which are lighter than GFRP and have excellent mechanical properties, are chiefly applied as materials for leisure ships such as expensive cruise boats. Since the hull's weight reduction directly affects their fuel efficiency, lightweight materials such as aluminum, GFRP and CFRP are representative lightweight materials for the ships.[1] CFRP is applied as a core material for military ships and leisure ships, and there is a case where a commercial ship propeller using CFRP was developed and applied in Japan.[2]

In this study, CFRP's lightweight structural design was carried out by replacing the GFRP leisure boat. The automated manufacturing process employing automated tape laying (ATL) replaces the existing hand lay-up as the hull manufacturing process is considered to increase productivity and reduce the manufacturing cost of ship manufacturing. This study aims to derive an optimal composite structure that can minimize the hull's weight and improve the CFRP material and ATL process's structural characteristics.

2. Structural Design of CFRP Boat

In this study, three types of lightweight leisure boat hulls were structurally designed. First, the baseline model with GFRP was designed, while two types of CFRP models were fabricated. Design specifications of the three types of model are shown in Fig. 1. In all three models, a GFRP or

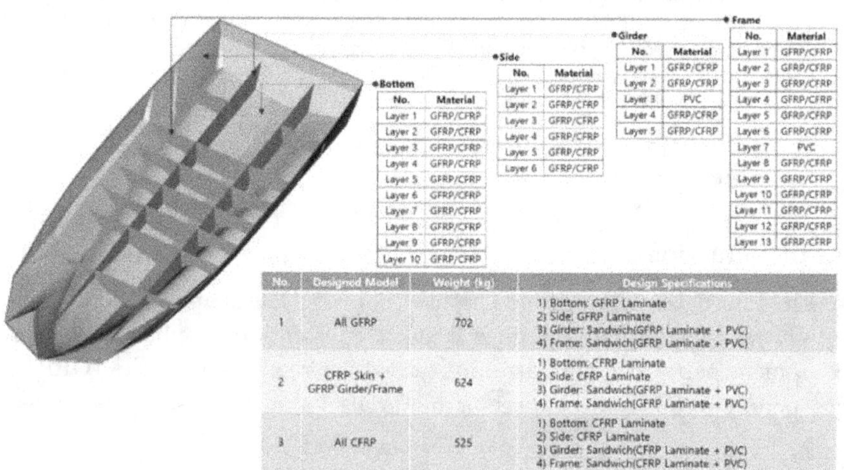

Fig. 1. (Color online) Design specifications of newly designed boat models.

CFRP laminate was applied to the skin and side parts, whereas sandwich structures using CFRP/GFRP laminate and PVC core were applied to the girder and frame parts of the boat designs. First, the standard model of the GFRP was designed with a total weight of 702 kg. The "CFRP Skin + GFRP Girder/Frame" model, in which CFRP is applied only to the skin area, was designed with a total weight of 624 kg, which is about 10% lighter than the standard model. Finally, the CFRP model was designed with a total weight of 525 kg, which is about 25% lighter than the baseline model.

3. Load and Boundary Conditions

For the structural analysis of the newly designed boats, loads and boundary conditions were applied to the boats' finite element model, as shown in Fig. 2. The pressure for evaluating the integrity of the structure was calculated by ISO standard. Mass in maximum load condition was applied to the pressure calculation for structural analysis, and m_{LDC} included the hull's weight, the weight of equipment such as engines, and the passengers' weight. As shown in the figure, a pressure of about 0.05 MPa was applied to the finite element model, and structural analysis for the three design specifications was carried out.

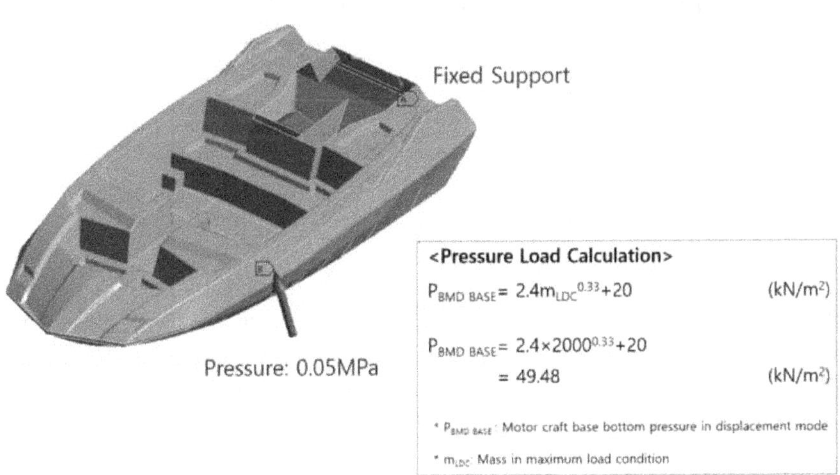

Fig. 2. (Color online) Load calculation and applied load of the structural analysis.

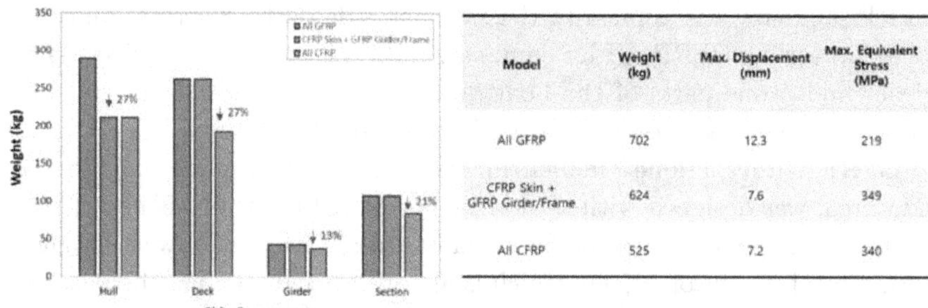

Fig. 3.   (Color online) Comparison of displacement for boat models.

4. Structural Analysis of CFRP Boat

Structural analysis for three kinds of lightweight boat models designed through this study was conducted, and the results are summarized in Fig. 3. As illustrated in the figure, the two models to which CFRP was applied instead of GFRP showed that the deformation was reduced by more than 35% under the same load condition compared to the existing GFRP ship model. This result occurred because the CFRP applied in this study has more than three times higher elastic modulus in the fiber direction than GFRP utilizing the hand lay-up process. Furthermore, the maximum equivalent stress of the two models to which CFRP was applied was increased by more than about 50% compared to the baseline model. It could be confirmed that the two newly designed models have sufficient structural integrity compared with the baseline model because the stress limit of CFRP is twice higher than that of GFRP.

5. Conclusions

In this study, the structural designs of an 18 ft class boat with composite materials were made.

First, a baseline model by GFRP was designed, and a structural design of about 700 kg in total weight was completed. Moreover, the model to which CFRP was applied only to the hull's skin, and the model to which CFRP was applied to the hull, girder, and frame were structurally designed and crafted to be about 10% and 25% lighter than the baseline model, respectively.

The structural integrity of the newly designed lightweight boat model was evaluated via Finite Element Analysis (FEA). It was confirmed that the two design models to which CFRP was applied under the load condition according to the ISO standard had sufficient structural integrity compared to the baseline model applying GFRP.

Through this study, the CFRP boat design for the ATL process was in progress to improve productivity and reduce weight compared to existing leisure boats. It was confirmed that the newly designed CFRP boat model reduced weight and improved structural characteristics compared to the GFRP boat model using hand lay-up.

References

1. A. Alikhani and R. Nedoushan, *Funct. Compos. Struct.* **2**(3), (2020).
2. B. S. Lee and W. R. Yu, *Funct. Compos. Struct.* **2**(1), (2020).

Laminated design for the application of composite materials for ship radar mast*

Sung Won Yoon[†] and Chang Wook Park[‡]

*Research Institute of Medium & Small Shipbuilding,
38-6, Noksansandan 232, Kangseo-gu, Busan 46757, Republic of Korea*
[†] *swyoon@rims.re.kr*
[‡] *cwpark@rims.re.kr*

The purpose of this study is to determine the correct estimation of laminated design for composite materials applied to a ship radar mast. Recently, as the IMO's environmental regulations have been strengthened to increase the energy efficiency of ships, the IMO has also begun to consider operational economics such as energy reduction through lightening the hull. Demand for lighter weight technology using composite materials is increasing. Examples would include lightweight large structures using composite materials, composite materials replacing metal design parts, and polymer composite materials applicable to marine environments. Therefore, in this study, the properties of the material were analyzed using a simulation program to verify the applicability of the composite material for ship radar mast. In addition, the reliability of the simulation result was secured through Fiber/Matrix/Ply correction and the calculation of the laminate dynamics. The results provide basic properties ($E33$, $G23$, $G13$, $\nu23$, $\nu13$, $S13$, $S23$) for computer analysis other than mechanical properties derived experimentally.

Keywords: Fiber reinforced plastic; laminated design; super structure; simulation program; radar mast.

1. Introduction

In this study, the stacking pattern analysis of composite materials for application to ship radar masts was performed. In composite materials, lamination design is very important because mechanical properties differ

---

[‡]Corresponding author.
*To cite this article, please refer to its earlier version published in the *International Journal of Modern Physics B*, Volume 35, 2140039 (2021), DOI: 10.1142/S0217979221400397.

depending on the lamination pattern of the reinforcement. A computer simulation program was used to predict the change in mechanical properties of the composite material according to the lamination pattern. It is very difficult to theoretically calculate the mechanical properties according to the lamination pattern of the reinforcement, and thus the product design direction varies according to the data obtained through experiments. However, since measuring mechanical properties through experiments is time-consuming and expensive, obtaining data through a simulation program is the fastest method. The lamination pattern is an important factor in determining the strength and longevity of the final structure. Therefore, the optimal stacking angle and thickness of the composite radar mast are examined using a simulation program. In addition, an optimal design method was derived through the presentation of the stacking pattern.[1-6]

2. Material Properties for Laminated Design

The mechanical properties of Fiber Reinforced Plastic (FRP) were evaluated for laminating pattern analysis of a radar mast. In this study, specimens were prepared using VaRTM method and cured in an oven at 16 h, 60°. The mechanical properties of specimens were measured by tensile, compression and shear strength tests using a universal test machine according to the ASTM standards. The properties of the FRP laminates for simulation are shown in Table 1.

In composite laminate, each layer has a different stacking angle, so both strength and stiffness are different for the same external load. The force

Table 1.   The mechanical properties of FRP laminates for simulation.

Description	GFRP		CFRP	
	Initial	Effective	Initial	Effective
(E11) Longitudinal modulus	37,700	37,700	121,000	121,000
(E22) Transverse modulus	4400	10,679	8360	8360
(G12) Shear modulus	5900	5900	6400	6400
(NU12) Poisson's ratio	0.28	0.28	0.28	0.28
(S11T) Longitudinal tensile strength	867	867	2.503	2.503
(S11C) Longitudinal compressive strength	551	551	633	633
(S22T) Transverse tensile strength	78	78	50	50
(S22C) Transverse compressive strength	110	110	140	140
(S12S) In-plane shear strength	53	53	64	64

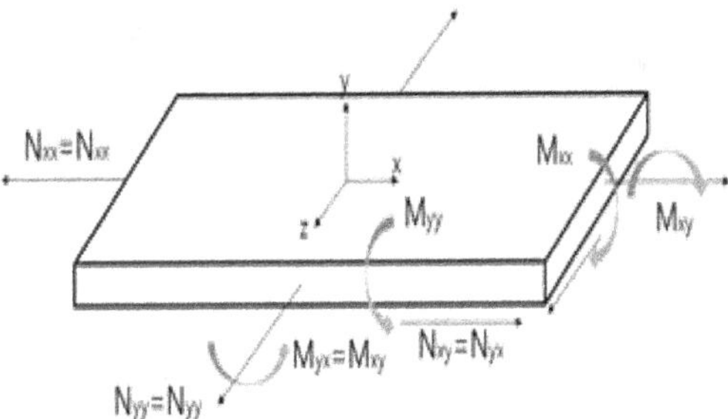

Fig. 1. Resultant forces and moments.

and moment applied to the entire laminate can be obtained by integrating different forces and moments for each layer according to the thickness of the laminate. As shown in Fig. 1, Nij is the force applied per unit length of the laminated plate (N/m), and Mij is the moment applied per unit length of the laminated plate (N/m^2).

3. Results and Discussions

In this study, the laminated design was analyzed using the MCQ-Composites program based on the basic physical property test data of fiber reinforced composite materials.

In this process, it is possible to secure mechanical properties in the vertical direction of the laminate, which is difficult to deduce through coupon test. The mechanical properties calculated in the Fiber/Matrix/Ply Calibration function are corrected values by reflecting the correlation between resin and fiber in the initially entered experimental value.

3.1. *Laminate mechanics*

Based on the results calculated by Fiber/Matrix/Ply Calibration, the composite material is reconstructed to recalculate the physical properties of the composite material. The laminated patterns of Glass-FRP (GFRP) are [0, 90, 45, −45] s, and the laminated patterns of Carbon-FRP (CFRP) are set to [0, 90] s. The final objective is [0, 90, 45, −45, 0, 90] s, which is a combination of GFRP and CFRP. The axial tensile stress (SxxT) was 634 MPa, while the compressive stress (SxxC) 294 MPa and shear stress (SxyS) were 96 MPa. Figure 2 shows the performance results of laminate mechanics.

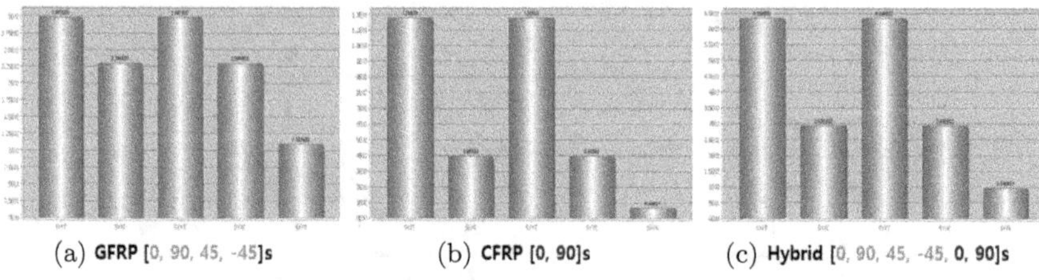

(a) GFRP [0, 90, 45, -45]s   (b) CFRP [0, 90]s   (c) Hybrid [0, 90, 45, -45, 0, 90]s

Fig. 2.   (Color online) Laminate mechanics: (a) GFRP, (b) CFRP, (c) hybrid.

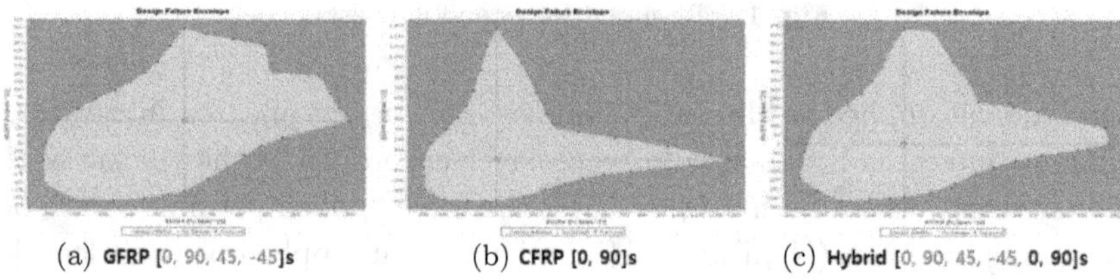

(a) GFRP [0, 90, 45, -45]s   (b) CFRP [0, 90]s   (c) Hybrid [0, 90, 45, -45, 0, 90]s

Fig. 3.   (Color online) Progressive failure: (a) GFRP, (b) CFRP, (c) hybrid.

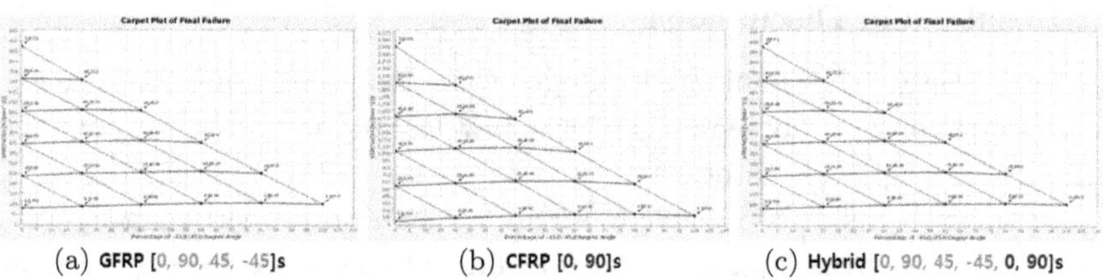

(a) GFRP [0, 90, 45, -45]s   (b) CFRP [0, 90]s   (c) Hybrid [0, 90, 45, -45, 0, 90]s

Fig. 4.   (Color online) Design failure envelope: (a) GFRP, (b) CFRP, (c) hybrid.

3.2. *Progressive failure*

Progressive failures were checked in the laminated design and damage prediction data were derived for each lamination direction. Axial compressive stress and tensile stress were partially generated, but normal stress was not affected. Figure 3 shows Ply's damage data.

3.3. *Design failure envelope*

When multiple times loads were applied to the composite material, the failure zone of the lamina or laminate was visually checked according to the failure criterion determined by the user. Figure 4 shows the failure criterion. From this figure, it was confirmed that the tensile stress of the

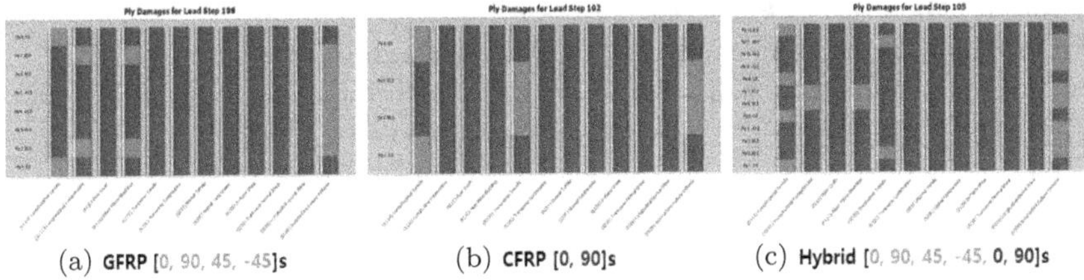

(a) **GFRP** [0, 90, 45, -45]s   (b) **CFRP** [0, 90]s   (c) **Hybrid** [0, 90, 45, -45, 0, 90]s

Fig. 5.   (Color online) Parametric carpet plots: (a) GFRP, (b) CFRP, (c) hybrid.

hybrid composite material starts at 225 MPa and the compressive stress starts at 190 MPa.

3.4. *Parametric carpet plots*

Parametric carpet plot was performed to intuitively understand the mechanical properties of the laminate according to the lamination angle. Figure 5 shows the results of Parametric carpet plots. From this data, it can be seen that stacking 0° (35%), 45° (30%) and 90° (35%) satisfies the target strength of 250 MPa.

4. Conclusions

In this study, the possibility of application of the ship radar mast was investigated through the analysis of the laminated design of composite materials, which are lightweight materials for improving the operational efficiency of ships. By reverse engineering the mechanical properties, a laminated design was implemented that satisfies the performance required for the radar mast. In addition, the reliability of the simulation result was secured through Fiber/Matrix/Ply correction and the calculation of the laminate dynamics.

Acknowledgments

This work was supported by the Technology Innovation Program (20010950, Development of Large-Scale Equipment for Eco-Ships) funded by the Ministry of Trade, Industry & Energy (MOTIE, Korea).

References

1. S. W. Yoon *et al.*, *Int. J. Mod. Phys. B* **32**, 1840067 (2018).
2. J. H. Cho *et al.*, *Int. J. Mod. Phys. B* **32**, 1840048 (2018).

3. O. A. Noqta *et al.*, *Funct. Compos. Struct.* **2**, 10793 (2020)
4. D. Parshuram and M. Sunil, *Int. J. Eng. Sci.* **2**, 160 (2013).
5. H. S. Noh and S. W. Koh, *Bull. Kor. Soc. Fisheries Technol.* **28**, 295 (1992).
6. H. J. Hwang *et al.*, *Funct. Compos. Struct.* **2**, 112483 (2020).

Fabrication and evaluation of CA-doped SrTiO$_3$ thermoelectric materials by molten salt method*

Kei-Ichiro Murai[†], Takuya Nishiura, Ryutaro Nagata and Toshihiro Moriga

*Graduate School of Technology, Industrial Social Science,
Tokushima University, 2-1 Minami-josanjima,
Tokushima, 770-8506, Japan*
[†] *keimruai@tokushima-u.ac.jp*

In this study, we focused on the molten salt method and attempted a simple synthesis of SrTi$_{0.8}$Co$_{0.2}$O$_3$. It was clarified that SrTi$_{0.8}$Co$_{0.2}$O$_3$ can be obtained under relatively mild conditions by using the molten salt. In all samples, the electrical conductivity increased with increasing temperature. Compared with the sample without the molten salt, the electric conductivity of the samples with the molten salt was greatly improved. The sample using the KCl-NaCl mixed salt showed the highest electrical conductivity of 83 S/cm at 973 K. The relative densities of all the samples using the molten salt were above 85%. It is considered that the improvement in the electrical conductivity is partly due to the increase in the relative density.

Keywords: Molten salt method; Ca-doped SrTiO$_3$; thermoelectric material.

1. Introduction

Thermoelectric conversion materials are considered as new energy conversion devices because they can convert heat into electricity and do not emit greenhouse gases. The thermoelectric properties are evaluated by the dimensionless figure of merit ZT $(= S^2\sigma T/\kappa)$. Here, S is the Seebeck coefficient, σ is the electrical conductivity, T is the absolute temperature, and κ is the thermal conductivity. Strontium titanate SrTiO$_3$ is one of the environmentally friendly thermoelectric conversion materials because of its

---

[†]Corresponding author.
*To cite this article, please refer to its earlier version published in the *International Journal of Modern Physics B*, Volume 35, 2140040 (2021), DOI: 10.1142/S0217979221400403.

low toxicity and excellent thermal stability.[1-9] However, there are problems such as the electric conductivity being close to zero and the thermal conductivity being high. Therefore, it is indispensable to improve the thermoelectric characteristics by introducing carriers. In these regards, the carrier concentration and carrier mobility play important role in improving the thermoelectric properties of $SrTiO_3$, but there are relatively many research reports on carrier concentration and very few research reports on carrier mobility in the current situation. Therefore, by focusing on carrier mobility and investigating the effects on electrical conductivity, Seebeck coefficient, we aimed to clarify the feasibility of new improvements in thermoelectric properties. It is necessary to repeat high-temperature firing and pulverization and mixing to obtain high-quality $SrTiO_3$ by the solid-phase reaction method. In this study, we focused on the molten salt method and attempted a simple synthesis of $SrTi_{0.8}Co_{0.2}O_3$. In addition, the effects of changing the type of salt used in the molten salt method on the relative density and thermoelectric properties of $SrTi_{0.8}Co_{0.2}O_3$ were investigated.

2. Experimental

The sample of $SrTi_{0.8}Co_{0.2}O_3$ was synthesized by the molten salt method. $SrCO_3$, TiO_2 (rutile type) and Co_3O_4 are used as starting materials. KCl, NaCl and a mixed salt with a weight ratio of KCl to NaCl of 1:1 are used as salts and each sample was named as Sample 2, Sample 3 and Sample 4, respectively. For comparison, a salt-free sample (Sample 1) was also synthesized under the same conditions. The starting materials were added to ethanol, mixed in a ball mill for 24 h, and dried overnight in a dryer to obtain a powder. A salt weighed so that the weight ratio of the starting material and the salt was 1:1 was added to this powder, and after pulverizing and mixing for 30 min, the mixture in an alumina crucible was calcined at 1200°C and 5 h. After distilled water was added to the obtained powder, salts were removed by filtration, and the mixture was dried overnight in a dryer to pressed into pellets. The pellet on an alumina boat was sintered under the conditions of 1400°C and 5 h to obtain the desired product. Phase identification-were carried out by powder X-ray diffraction (XRD) (Rigaku SmartLab, Japan). To measure electrical conductivity and Seebeck coefficient, the samples were cut down from the sintered pellets into rectangular bars with a dimension of 5 mm × 5 mm × 10 mm. The electrical conductivity was measured by the two-probe method using Pt wires and Pt electrodes. The Seebeck coefficient was calculated from thermoelectric voltage generated between both ends of a sample ΔV and temperature difference between the two ends ΔT. Both properties were measured in air in the temperature range from 573 K to 973 K. Thermal conductivity was measured by laser flash method (Ulvac TC-7000).

3. Results and Discussion

From the XRD spectra in Fig. 1 of each sample before sintering at 1400°C, a slight unreacted starting material was confirmed in Sample 1. On the other hand, in Samples 2–4, no starting material was confirmed, and only SrTiO₃ was confirmed. It was clarified that the target product can be obtained under relatively mild conditions by using salt. After sintering at 1400°C, it was confirmed that all the samples including Sample 1 had a single phase. Figure 2 shows the electrical conductivity of each sample after sintering at 1400°C and 5 h. In all the samples, the electric conductivity increased with increasing temperature and showed semiconductor-like

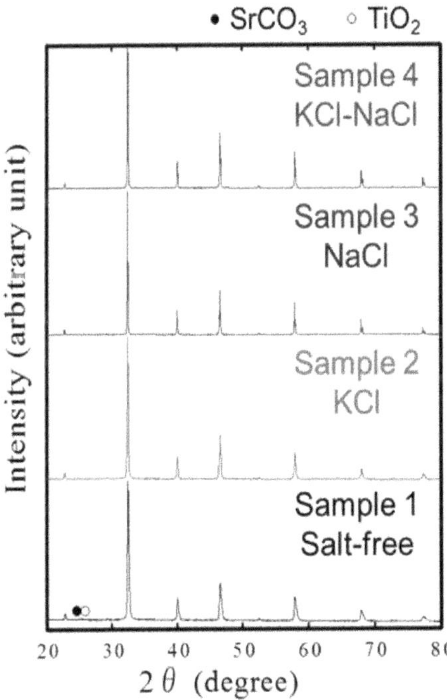

Fig. 1. (Color online) XRD patterns of SrTi₀.₈Co₀.₂O₃ samples before sintering at 1400°C, 5 h.

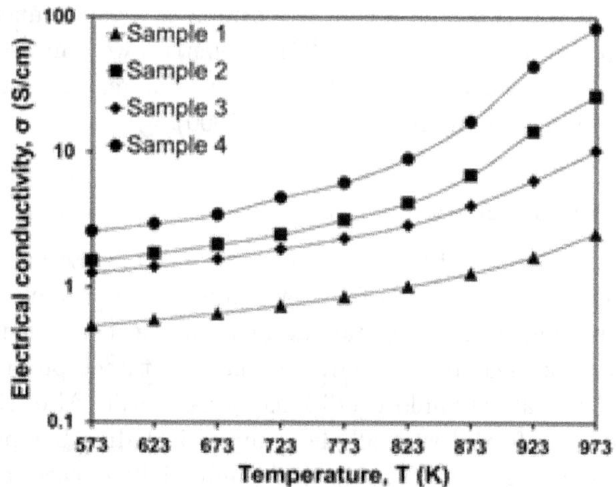

Fig. 2.   Electrical conductivities of $SrTi_{0.8}Co_{0.2}O_3$ after sintering at 1400°C, 5 h.

behavior. Compared with Sample 1 without salt, the electrical conductivity of Samples 2–4 with salt was significantly improved. The electrical conductivity of Samples 1–4 showed the maximum value at 973 K, which was 2.5 S/cm, 26 S/cm, 10 S/cm and 83 S/cm, respectively. Sample 4 showed the highest electrical conductivity. The electrical conductivity σ is given by the following equation.

$$\sigma = e \times n \times \mu, \tag{1}$$

where e is an elementary charge, n is a carrier concentration, and μ is a carrier mobility. The electrical conductivity is proportional to the carrier mobility. The relative densities of Samples 1–4 were 76%, 87%, 85% and 90%, respectively. In general, it is considered that the higher the relative density value, the better the carrier mobility, because the particles are denser as the relative density value is larger. Since the relative density was significantly improved in the salt-based sample, it is considered that the improvement in the electrical conductivity is partly due to the improvement in carrier mobility with the increase in relative density. In the sample using salt, the electric conductivity increases in the order of Samples 3–4, and the relative density also increases in this order. Figure 3 shows the results of thermal conductivity measurement of each sample after sintering at 1400°C and 5 h. In all samples, the thermal conductivity increased with

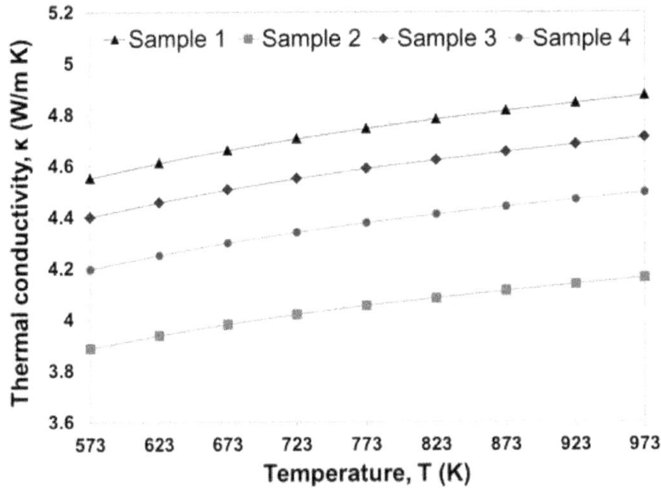

Fig. 3. (Color online) Thermal conductivities of SrTi$_{0.8}$Co$_{0.2}$O$_3$ after sintering at 1400°C, 5 h.

increasing temperature. The thermal conductivity of Samples 1–4 showed the minimum value at 573 K, which was 4.6 W/mK, 3.9 W/mK, 4.4 W/mK and 4.2 W/mK, respectively.

Of all the samples, Sample 2 showed a relatively low thermal conductivity. It is thought that phonon scattering, which is a factor in reducing thermal conductivity, is likely to occur as the average particle size decreases, the surface area of the grain boundaries increases. The decrease in thermal conductivity of Sample 2 is thought to be due to the relatively small average particle size.

Figure 4 shows the ZT value of each sample after sintering at 1400°C and 5 h. In all samples, the ZT value increased with increasing temperature. Compared with Sample 1 without salt, it was clarified that the ZT values of Samples 2–4 with salt were significantly improved. The ZT values of Samples 1–4 showed the maximum values at 973 K, which were 0.0023, 0.023, 0.0090 and 0.062, respectively. The reason why the ZT value of the salt-based sample was significantly improved was that the electrical conductivity was significantly improved, the decrease in the Seebeck coefficient was suppressed and the thermal conductivity was relatively small. Therefore, it was clarified that the molten salt method is a useful synthetic method for improving the ZT value as compared with the solid phase reaction method.

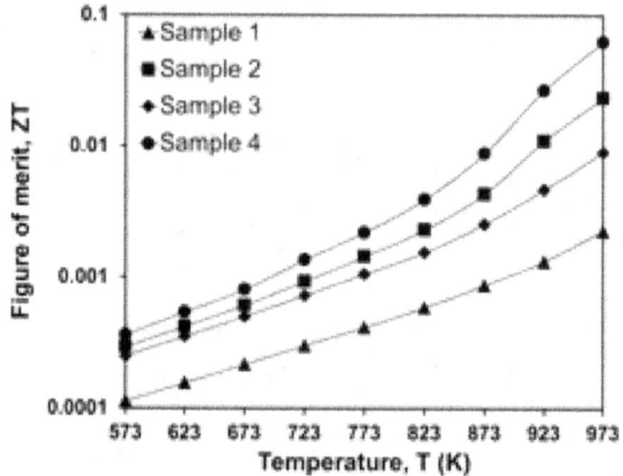

Fig. 4.   Figure of merits of $SrTi_{0.8}Co_{0.2}O_3$ before sintering at 1400°C, 5 h.

4. Conclusion

The sample using the KCl-NaCl mixed salt showed the highest electrical conductivity of 83 S/cm at 973 K and the highest ZT of 0.062. The relative densities of all the samples using the molten salt were above 85%. It was clarified that the molten salt method is a useful synthetic method for improving the ZT value as compared with the solid phase reaction method.

References

1. S. Ohta *et al.*, *J. Appl. Phys.* **97**, 034106 (2005).
2. H. Obara, A. Yamamoto *et al.*, *Jpn. L. Appl. Phys.* **43**, L540 (2004).
3. T. Okuda *et al.*, *Phys. Rev. B* **63**, 113104 (2001).
4. S. Ohta, T. Nomura, H. Ohta *et al.*, *Appl. Phys. Lett.* **87**, 092108 (2005).
5. H. Muta, K. Kurosaki and S. Yamanaka, *J. Alloys Compd.* **350**, 292 (2003).
6. J. Wang, B.-Y. Zhang, H.-J. Kang *et al.*, *Nano Energy* **35**, 387 (2017).
7. K. Park, J. S. Son, S. I. Woo, K. Shin *et al.*, *J. Mater. Chem. A.* **2**, 4217 (2014).
8. T. Teranishi, Y. Ishikawa, H. Hayashi *et al.*, *J. Am. Ceram. Soc.* **96**, 2852 (2013).
9. D. Srivastava, C. Norman, F. Azough *et al.*, *J. Alloys Compd.* **731**, 723e73 (2018).

Applications of a novel detector for pencil beam scanning proton therapy beam quality assurance*

Pei-Ying Yang[†,‡], Yang-Wei Hsieh[†,‡], Chen-Lin Kang[‡], Chin-Dar Tseng[†], Chih-Hsueh Lin[†], Chin-Shiuh Shieh[†], Pei-Ju Chao[†,‡] and Tsair-Fwu Lee[†,‡,§,¶]

[†] *Medical Physics and Informatics Laboratory of Electronics Engineering,*
National Kaohsiung University of Science and Technology,
Kaohsiung 80778, Taiwan, ROC

[‡] *Department of Radiation Oncology,*
Kaohsiung Chang Gung Memorial Hospital
and Chang Gung University College of Medicine,
Kaohsiung 83342, Taiwan, ROC

[§] *Department of Medical Imaging and Radiological Sciences,*
Kaohsiung Medical University, Kaohsiung 80708, Taiwan, ROC
[¶] *tflee@nkust.edu.tw*

This study utilized a new type of detector, the CROSS II (Liverage Biomedical Inc., Taiwan), to perform a beam quality assurance (QA) procedure on a Sumitomo (Sumitomo Heavy Industries, Inc., Japan) pencil beam linear scanning proton therapy machine. The Cross II can monitor proton Pristine Bragg peak range, beam width, beam size, beam position, and scanning speed. All the data presented here were collected during a time span of over one year. The accuracy of the QA program could be verified if all the QA items were tested stably and within the programmed tolerances. Our results showed that the proton range remained within the ± 2 mm tolerance, with the majority of measurements within ± 0.5 mm, ± 2 mm for spot size, 1.5 mm for spot position, and $\pm 2\%$ for scanning speed. We found that the CROSS II detector is in high precise and steady state with highly efficient. Our proton therapy system was also proven to be in an accurate and reliable condition according to our QA results.

Keywords: Proton; Pristine Bragg peak; detector; pencil beam scanning; quality assurance.

[¶]Corresponding author.
*To cite this article, please refer to its earlier version published in the *International Journal of Modern Physics B*, Volume 35, 2140041 (2021), DOI: 10.1142/S0217979221400415.

1. Introduction

Pencil beam scanning (PBS) is now in the mainstream of proton therapy. Such a dedicated and complex beam delivery technique can provide a more conformal dose in target coverage and a lower dose in organs at risk (OARs) than double scattering (DS) proton system.[1] Nevertheless, proton range, spot size and position uncertainties might cause dose discrepancies during PBS delivery.[2] A quality assurance (QA) protocol that can monitor PBS characteristics should be implemented to guarantee an accurate, consistent and safe dose delivered to patients. According to a recommendation from AAPM TG-224,[3] each proton center should develop its own QA protocol based on its equipment design and settings.[4–6] QA should also be performed before starting the treatments each day. The aim of the beam QA program for the PBS proton beam is to assure the accuracy or constancy of the following beam delivery system components: range of the individual spots, spot depth dose distribution, spot lateral fluence and dose monitor calibration, dose profiles, spot positioning. One of the major challenges in proton QA checks is the time limitation to perform such procedures.

2. Materials and Methods

Our PBS system was constructed by the Sumitomo proton therapy system (Sumitomo Heavy Industries, Inc., Japan) in the end of 2018.[7] The objective of this work was to utilize the recently commercially available tools, such as the CROSS II (Liverage Biomedical Inc., Taiwan), to execute most of our PBS QA items. Beam characteristics can be monitored by the CROSS II. The daily QA is needed to verify: (1) proton Pristine Bragg peak range (range); (2) spot size; (3) spot position; (4) scanning speed aspects of the treatment machine before patient treatment. Our scanning beamline is capable of delivering 70 MeV to 230 MeV energies. Because it is nearly impossible the dosimetry features of proton PBS for all these energies daily, 150 MeV was used for our fast-daily QA program. Hence, we analyzed and present one year of 150 MeV QA results after implementing the mentioned device in our systems. This information can be of benefit to proton centers that are in need of developing a daily QA program. As shown in Figs. 1(a) and 1(b), the CROSS II parallel ionization chamber is excellent, with a convenient setup, high special resolution (0.2 mm), short response time for signal acquisition and large measurement area of 33×33 cm^2, and it can measure the scanning speed of the line scanning beam.

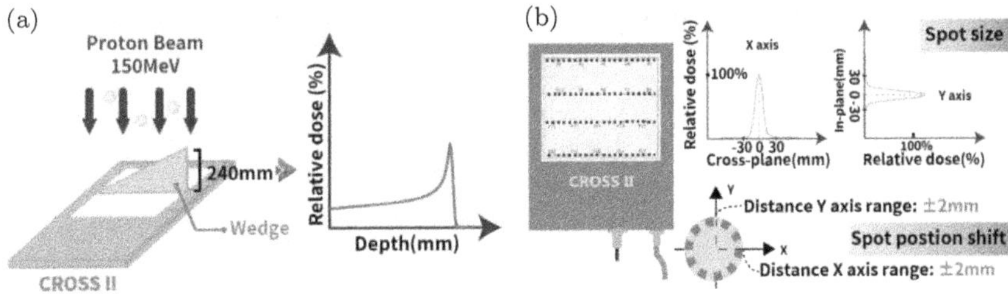

Fig. 1. (Color online) CROSS II used for verification of (a) percent depth dose, R_{80} and D_{90} — P_{90} width; (b) spot size and position accuracy for 150 MeV.

3. Results and Discussion

The CROSS II detector is highly efficient to complete the daily QA checks that takes about 10 min room time per gantry room for each energy to finish. Table 1 presents the QA 244 records for one year for the first treatment room PBM nozzle. We first did statistics and an error tolerance comparison. After that, we evaluated the long-term stability of each item. The QA data will show different beam parameters, even when using the same brand of proton machine. We ensured that each QA item was within its tolerance level. Figure 2 represents the variation of daily beam parameters for 150 MeV proton beams over a one-year period. Figure 2(a) shows that for the distal 80% dose the proton range was within ±0.5 mm. Figure 2(b) shows six scanning speed data: 1, 2, 5, 10, 15 and 20 mm/ms were measured. The long-term observation indicated that the deviation was about

Table 1. QA results for proton range constancy, spot size, scanning speed and position accuracy.

Beam parameters	Tolerance	Mean	Standard deviation
Proton range	±1 mm for $D_{80\%}$	−0.10 mm	0.15
	±2 mm for $DP_{90\%-90\%}$	−0.07 mm	0.05
Spot size	±1 mm	Y: 0.12 mm	Y: 0.18
		X: −0.11 mm	X: 0.14
Spot position	±1.5 mm	0.86 mm	0.26
Scanning speed			
1 mm/ms		0.01%	0.05%
2 mm/ms		0.05%	0.04%
5 mm/ms	±2%	−0.09%	0.06%
10 mm/ms		0.21%	0.06%
15 mm/ms		0.26%	0.06%
20 mm/ms		0.55%	0.36%

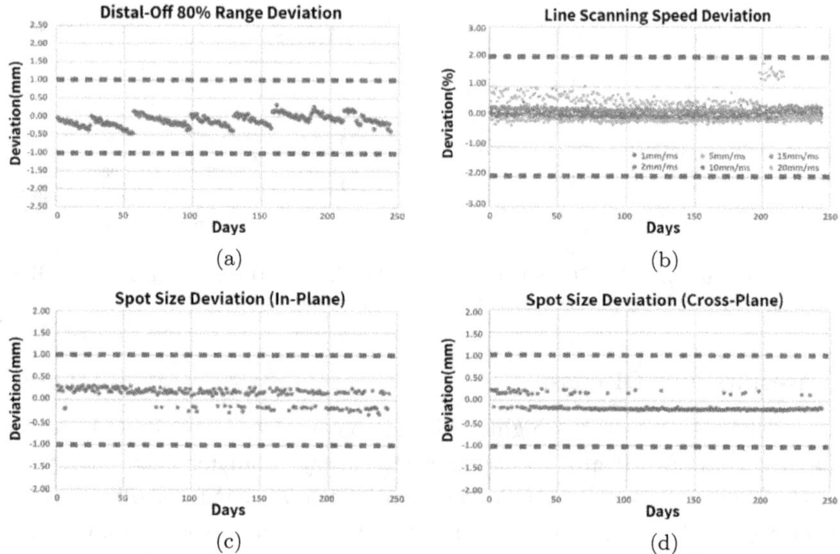

Fig. 2. (Color online) Variation of daily beam parameters for 150 MeV proton beams over a one-year period. The data represent (a) proton range; (b) scanning speed; (c) spot size (in-plane) and (d) spot size (cross-plane).

-0.3% to 0.5%. Nevertheless, the 20 mm/ms deviation was -0.11% to 1.8%. Figure 2(c) shows that the spot size in the in-plane direction was within ± 0.26 mm. Figure 2(d) shows that the spot size in the cross-plane direction was within ± 0.35 mm.

The daily QA program that we have developed and implemented in our proton center was designed to check with high accuracy for beam characteristics and equipment parameters on a daily basis. A comprehensive set of checks can be performed using a combination of a commercially available CROSS II, such as beam characteristics monitored accuracy excellent, minimize the time for complete PBS daily QA, light weight and easy installation. However, previous studies[8,9] showed that the uncertainties in proton range, spot size and spot position can bring the large dose change in treatment. To solve the uncertainties, a QA procedure should be established for each facility. A dedicated QA program helps to eliminate the potential risks of the proton therapy equipment before the patient is treated.

4. Conclusions

This study focused on the daily QA measurements of the PBS beam quality. We validated and verified the proton Pristine Bragg peak range, spot size, spot position and scanning speed. We found that the CROSS II detector is highly precise and in a steady state with highly efficient. Our proton therapy system was also proven to be in an accurate and reliable condition according to our daily QA results. The QA results also showed us the stability of the Sumitomo PBS nozzle.

Acknowledgments

This work was supported by the grants No. MOST 109-2221-E-992-011-MY2 and MOST 109-2623-E-992-006-NU.

References

1. X. Zhang *et al.*, *Int. J. Radiat. Oncol. Biol. Phys.* **77**, 357 (2010).
2. H. Paganetti, *Proton Beam Therapy* (IOP Publishing, Bristol, 2017).
3. B. Arjomandy *et al.*, *Med. Phys.* **46**, e678 (2019).
4. X. Ding *et al.*, *J. Appl. Clin. Med. Phys.* **14**, 115 (2013).
5. J. Lambert *et al.*, *J. Appl. Clin. Med. Phys.* **15**, 217 (2014).
6. S. Rana *et al.*, *J. Appl. Clin. Med. Phys.* **20**, 29 (2019).
7. R. Kohno *et al.*, *Int. J. Part. Ther.* **3**, 429 (2017).
8. H. Paganetti, *Phys. Med. Biol.* **57**, R99 (2012).
9. H. Li *et al.*, *Med. Phys.* **40**, 021703 (2013).

Multi-layer thin-film deposition for high-performance X-ray field-emission characteristics*

Tae Hwan Jang[†] and Tae Gyu Kim[‡]

*Department of Nanomechatronics Engineering,
Pusan National University, Busan 46241, Republic of Korea*
[†]*taehwan110@hanmail.net*
[‡]*tgkim@pusan.ac.kr*

Mun Ki Bae

*Department of Nanofusion Technology, Pusan National University,
Miryang, Gyeongnam 50463, Republic of Korea*
wck9230@naver.com

Kyuseok Kim[§] and Jaegu Choi[¶]

*Electro-Medical Device Research Center,
Korea Electrotechnology Research Institute (KERI) Ansan-si,
Gyeonggi-do 15588, Republic of Korea*
[§]*kskim502@keri.re.kr*
[¶]*jgchoi@keri.re.kr*

In this study, we developed a nanoscale emitter having a multi-layer thin-film nanostructure in an effort to maximize the field-emission effect with a low voltage difference. The emitter was a sapphire board on which tungsten–DLC multi-player thin film was deposited using PVD and CVD processes. This multi-layer thin-film emitter was examined in a high-vacuum X-ray tube system. Its field-emission efficiency according to the applied voltage was then analyzed.

Keywords: Field emission; DLC; multi-layer thin-film emitter.

[‡]Corresponding author.
*To cite this article, please refer to its earlier version published in the *International Journal of Modern Physics B*, Volume 35, 2140043 (2021), DOI: 10.1142/S0217979221400439.

1. Introduction

Electron emitters are field-emission (FE) sources applied to X-ray and cathode ray tube (CRT). In recent years, studies have been actively made to improve emitters' field-emission efficiencies. Fowler–Nordheim (F–N) theory states that most electron emissions come from nanomaterials.[1,2] The CNT has an excellent performance from the perspectives of emitted current. Besides, it has high aspect ratio, thermal conductivity and chemical stability. On the other hand, emitters with CNT have a problem in terms of consistent manufacturing process and period of life, which is short, leading to a low performance. Particularly, as cathode current in an X-ray tube determines image resolution requiring higher density of cathode current, high durability and reliability are important to commercialize its application. This paper suggests a new diamond-like carbon (DLC) multi-layer thin-film emitter for field emissions of X-ray tube.[3] This study manufactured a nanoscale emitter based on a multi-layer thin-film nanostructure to maximize field emissions at a low voltage difference. Tungsten (W)–DLC multi-layer thin film was also created to improve its field-emission efficiency. Diamond-like carbon is widely used in semiconductors, machine parts and optic materials because it has low friction coefficient and high insulation. When highly insulating DLC meets highly conductive tungsten and forms a multi-layer thin-film structure, a very clear boundary plays a role as a back-plate of a metal layer which emits current at the same time. The multi-layer thin-film board was made of sapphire, which was extremely firm and heat-resistant. Tungsten and DLC were then deposited on the board in multiple layers. Field emissions of the emitter were measured using X-ray tube. Field emissions according to applied voltages were also examined.

2. Experimental Procedures

2.1. *DLC–tungsten multi-layer thin-film-emitter deposition*

The tungsten layer was processed with a physical vapor deposition (PVD) system using a direct current (DC) power, while the DLC layer was made with a chemical vapor deposition (CVD) method. The size of the board was 10×10 mm^2 with 100 μm in thickness. In a pretreatment process, the sapphire board was cleansed in a mixed liquid of sulfuric acid and hydrogen peroxide (3:7) for 15 min. Then ultrasonic cleaning was done for the board using acetone, ethanol and distilled water in order. Ensuring that the board and the DLC film were stuck together, a buffer layer ($\leq$10 nm in thickness)

Table 1.   Deposition conditions for multi-layer thin films.

Layer(s)/material	Power type and voltage	Gas source (sccm)	Vacuum pressure (Torr)	Film thickness (nm)
1st/HMDSO	RF (450 W)	HMDSO (10)	2.0×10^{-2}	10
2nd/DLC	RF (450 W)	Ar (30) and C_2H_2(50)	2.0×10^{-2}	80
3rd, 5th and 7th/W	DC (1000 V)	Ar (50)	1.0×10^{-3}	60
4th, 6th and 8th/DLC	DC (400 V)	Ar (30) and C_2H_2(50)	2.0×10^{-2}	80

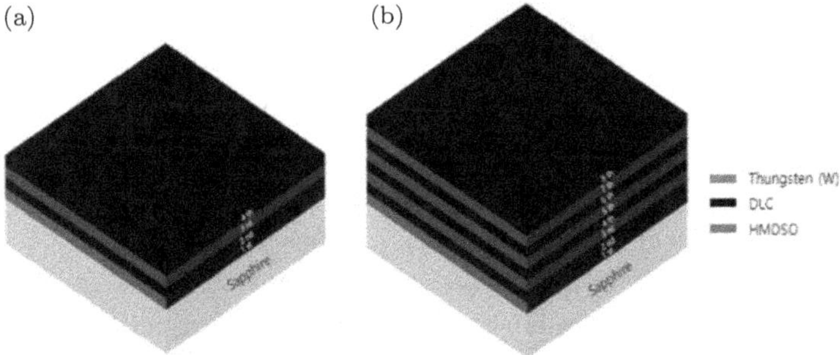

Fig. 1.   (Color online) Schematic of multi-layer thin films: (a) four-layer thin films and (b) eight-layer thin films.

was deposited following the CVD process which used hexamethyl disiloxane (HMDSO). The deposition conditions are shown in Table 1. Films are from one-layered to eight-layered. Figure 1 displays the schematic diagram.

2.2. *F–N theory and plot*

The FE phenomenon can explain quantum tunneling inside a metal as it exceeds Fermi energy by work function.[1] The FE phenomenon formula is shown as follows:

$$J = A\beta^2 V^2 \exp\left(\frac{-B}{\beta V}\phi^{\frac{3}{2}}\right) [A \cdot \mathrm{cm}^{-1}],$$

$$A = 1.5414 \times 10^{-6} \left[\frac{A \cdot \mathrm{eV}}{V^2}\right], \tag{1}$$

$$B = 6.8308 \times 10^{7} [\mathrm{eV}^{-\frac{3}{2}} \cdot V \cdot \mathrm{cm}^{-1}]$$

where J is current density, V is applied voltage influencing on the emitter surface and A and B are constants.[2] If the emitter is influenced by a high level of V or its surface curvature is small, the current density becomes big. As a method to check whether FE based on the F–N theory has been achieved, a curve of $\ln(1/V^2)$ versus $1/V$ is mainly used, and it is so-called the F–N plot. The multi-layer thin-film emitter was 10 μm wide and 3 mm high. The FE was measured using a two-electrode X-ray tube formed with anode target, cathode and vacuum-sealed glass bulb. The emitter was launched on the X-ray tube. The distance between the emitter and electrodes was set at 2 mm. The internal pressure was maintained between 10^{-6} Torr and 10^{-7} Torr to minimize the disturbances in field emission. Both four-layered and eight-layered emitters were used for experiments for FE under the same condition.

3. Results and Discussion

3.1. *Thin-film analysis*

In order to ensure that DLC and tungsten films were well coated by CVD and PVD, Raman and XRD analyses were performed. The results are shown in Fig. 2. According to Raman spectroscopy, DLC film has a mix of diamond structure (sp^3) and graphite structure (sp^2). As shown in Fig. 2(a), the DLC film is well located in which D-Peak (1350 cm^{-2}) and G-peak (1580 cm^{-2}) are blended. Figure 2(b) shows the XRD analysis results of the PVD tungsten film, revealing that the tungsten thin film is deposited on the sapphire board (1120) as seen in (200) and (220).

3.2. *SEM analysis*

Figure 3 shows the SEM images of four-layered and eight-layered tungsten–DLC cross-sections. It was found that the DLC film was 80 nm thick and the tungsten film was 60 nm thick.

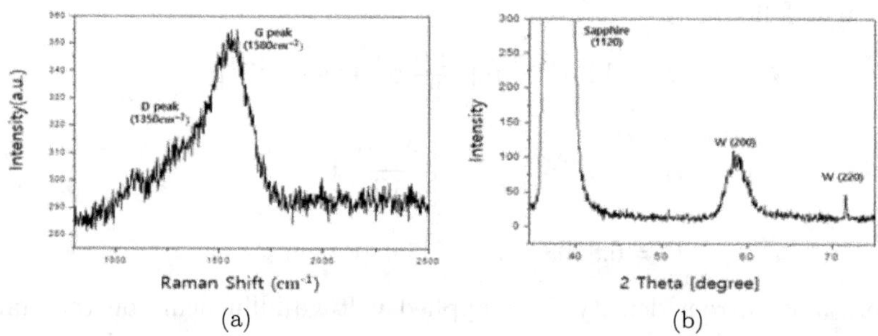

Fig. 2.   Raman result of DLC layer (a) and XRD result of tungsten layer (b).

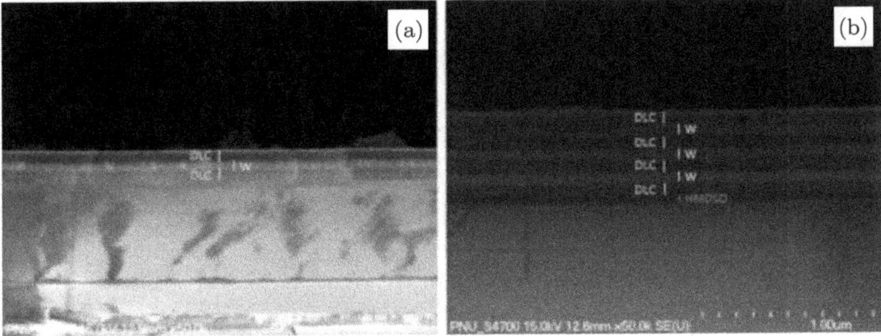

Fig. 3. (Color online) SEM images of multi-layer thin film: (a) four-layered and (b) eight-layered.

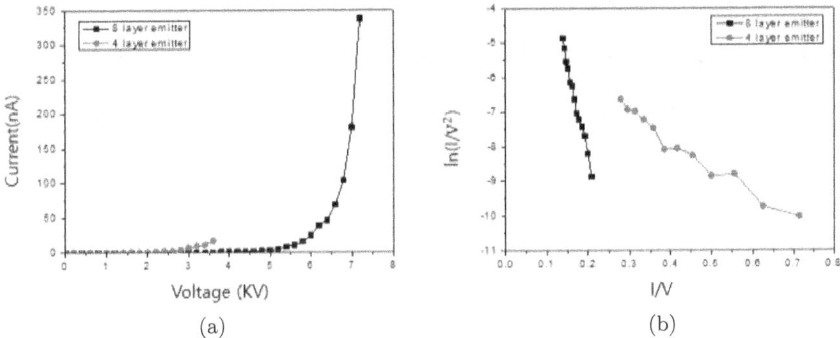

(a)   (b)

Fig. 4. (Color online) Current–voltage (I–V) curve and (b) F–N plot of multi-layer emitters.

3.3. *Field-emission analysis*

Figure 4(a) shows the measured results of emitted current for each voltage applied to four-layered and eight-layered emitters. The maximum current of the eight-layered emitter was 337 nA at 7.2 kV, while it was 17 nA at 3.6 kV for the four-layered one. This represents that the more layers the emitter has, the higher the initial voltage for the current it has from the beginning. Therefore, increasing the number of layers from four to eight produces a higher current than an emitter with fewer layers since it can operate at higher applied voltages. Figure 4(b) shows the F–N plot of Fig. 4(a). It reveals that the eight-layered emitter has a better efficiency as per the F–N formula than the four-layered emitter.

4. Conclusion

In this study, multi-layer thin films were deposited structured with the formation of DLC and tungsten layers, respectively, on a sapphire substrate in order to fabricate an FE emitter. SEM and Raman analyses confirmed that multi-layer thin films were well coated and a multi-layer thin-film DLC emitter was produced. In X-ray field-emission calculation, the eight-layered emitter recorded 337 nA at 7.2 kV, which was the maximum current, whereas the four-layered emitter recorded 17 nA at 3.6 kV. The F–N plot assured that the eight-layered emitter was better in terms of field-emission efficiency.

References

1. R. H. Fowler and L. Nordheim, *Proc. R. Soc. Lond. A, Contain. Pap. Math. Phys. Charact.* **119**, 173 (1928).
2. J. A. Nation, L. Schachter, F. M. Mako, L. K. Len, W. Peter, C.-M. Tang and T. Srinivasan-Rao, *Proc. IEEE* **5**, 865 (1999).
3. K. Kim, D.-G. Kang, T. H. Jang, T. G. Kim and J. Choi, *Jpn. J. Appl. Phys.* **34**, 1623 (2020).

Investigation of electrical properties of peald-deposited Ti/Al$_2$O$_3$/Al/Si MIM capacitors*

Sumit Patil, Viral Barhate and Ashok Mahajan[†]

*Department of Electronics, School of Physical Sciences,
Kavayitri Bahinabai Chaudhari North Maharashtra University,
Jalgaon 425001, Maharashtra, India*
[†]*ammahajan@nmu.ac.in*

Haoyu Xu, Mohammad Rasadujjaman and Jing Zhang

*School of Information Science and Technology,
North China University of Technology, Beijing 100144, P. R. China*

MIM devices fabricated with 10-nm thickness of Al$_2$O$_3$ high-k thin film deposited using plasma-enhanced atomic layer deposition (PEALD) system on Al-coated Si substrate were investigated. The structural, morphological and electrical properties of Ti/Al$_2$O$_3$/Al/Si MIM capacitors as-deposited and post-deposition annealed (PDA) at different temperatures were studied and compared. Al$_2$O$_3$ thin films were investigated using atomic force microscopy (AFM) and X-ray diffraction (XRD) and Ti/Al$_2$O$_3$/Al/Si MIM capacitors were characterized by current–voltage (I–V) and capacitance–voltage (C–V) measurements. The stable phase formation of Ti/Al$_2$O$_3$/Al/Si MIM capacitor provides the lowest leakage current density in the range of nA/cm^2 for as-deposited and annealed films.

Keywords: High-k thin films; post-deposition annealing; MIM capacitor; dielectric thin films.

1. Introduction

Al$_2$O$_3$ has attracted great attention as an alternative dielectric because of its highest bandgap ($\sim$8.9 eV) and resistance, compatibility on Si and thermal stability upto 1000°C.[1-4] Consequently, the conventional SiO$_2$ gate

[†]Corresponding author.
*To cite this article, please refer to its earlier version published in the *International Journal of Modern Physics B*, Volume 35, 2140045 (2021), DOI: 10.1142/S0217979221400452.

dielectrics were replaced by the high-k dielectric oxides (e.g., pure Al_2O_3) to meet the requirements by sustaining a lower leakage current.[5,6] Efforts have also been made for the fabrication of high-k thin films as an insulating layer in next-generation MIM devices that have forced the reduction of device dimensions towards nanometer scale requiring fully dense devices with higher speed and low power consumption. The high-k thin-film MIM capacitors are used in various applications like in deep trench DRAM technology, RF circuits and integrated circuits.[1,4,7] However, the final characteristics of high-k thin films depend on their thicknesses and may also depend on the deposition temperature at which the thin films are applied.

In this work, the structural, morphological and electrical characterizations of as-deposited and post-deposition annealed (PDA) Al_2O_3 high-k dielectric thin-film-based MIM capacitors have been carried out and their structural and electrical properties are compared. Section 2 describes experimental details, results are discussed in Sec. 3 and Sec. 4 concludes the paper.

2. Experimental Details

First, aluminum deposition with a thickness of 70 nm as the bottom electrode was done on pre-cleaned Si substrates by using the plasma-enhanced atomic layer deposition (PEALD) system. Further, the Al_2O_3 high-k thin films were deposited on Al-coated Si substrate with O_2 plasma by using the same PEALD system with the precursor trimethylaluminum (TMA)[8,9] for aluminum at 120°C for 10-nm thickness at a constant substrate temperature of 200°C and an RF plasma power of 60 W. Thereafter, post-deposition annealing in muffle furnace was employed on Al_2O_3/Al/Si high-k thin films followed by the deposition of Ti top metal electrode by using RF sputtering. After this experiment, surface properties were studied by using atomic force microscopy (AFM), while the structural properties were investigated by X-ray diffraction (XRD) and the electrical characterizations have been carried out using Keithly setup for current–voltage (I–V) and capacitance–voltage (C–V) measurements.

3. Results and Discussion

3.1. *X-ray diffraction study*

Figure 1 shows the X-ray diffraction patterns for the as-deposited and post-deposition annealed Al_2O_3 thin films at various temperatures. Step size of

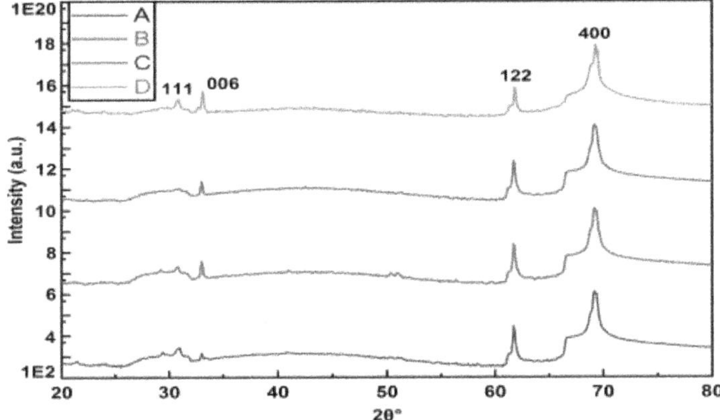

Fig. 1. (Color online) XRD patterns of as-deposited (curve A), 250°C (curve B), 300°C (curve C) and 350°C (curve D) annealed Al$_2$O$_3$/Al/Si thin films.

0.1 and $\lambda = 1.54060$ Å with the scan range of 2θ from 20° to 80° were used to perform the measurements. The XRD (MiniFlex 600 Rigaku, Japan) spectra of as-deposited and annealed films depict two sharp peaks at around 61.7° and 69.2° corresponding to the (122) phase of Al$_2$O$_3$ with single crystalline structure and cubic structure of Si having (400) phase, respectively. The peaks assigned based on the JCPDS Data File No. 85-1337 show the rhombohedral structure of Al$_2$O$_3$. All films show sharp diffraction peak at around 33° corresponding to the formation of hexagonal Al$_2$O$_3$ with (006) phase and resembling the JCPDS File No. 21-0010. While the additional peak at ~31° of Al$_2$O$_3$ films shows the orthogonal-structured Al$_2$SiO$_5$ composite formation at the Al–Al$_2$O$_3$ interface having (002) phase according to the JCPDS File No. 10-0369. In this work, the film thickness of around ~10 nm gives the change in phases even at lower temperatures.

3.2. *Atomic force microscopy study*

The surface roughness of Al$_2$O$_3$ high-k thin films over Al-coated Si was investigated by AFM (TriA 100 SPM, APER, Italy). The 3D AFM micrograph of the Al$_2$O$_3$/Al/Si thin film annealed at 300°C is shown in Fig. 2 over the scanning area of 100 nm × 100 nm, respectively. It is observed that as the PDA temperature increases, the root-mean-square roughness (Sq) decreases (~1.9 nm). The 3D image shows the smooth and crack-free distribution of the grains over the entire surface for Al$_2$O$_3$ thin films[10] in Fig. 2.

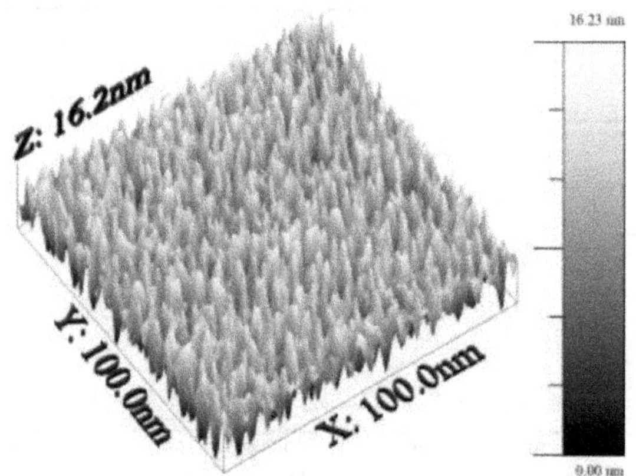

Fig. 2.   (Color online) The 3D AFM image of 300°C annealed Al_2O_3/Al/Si thin film.

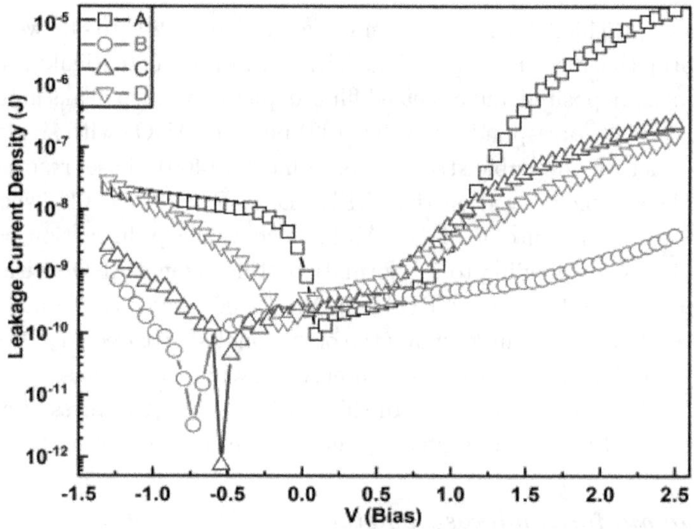

Fig. 3.   (Color online) J–V characteristics of as-deposited (curve A), annealed at 250°C (curve B), 300°C (curve C) and 350°C (curve D) Ti/Al_2O_3/Al/Si MIM capacitors.

3.3. *Current–voltage characteristics*

Figure 3 shows the leakage current density characteristics of Ti/Al_2O_3/Al/Si thin films as-deposited and annealed at different PDA temperatures

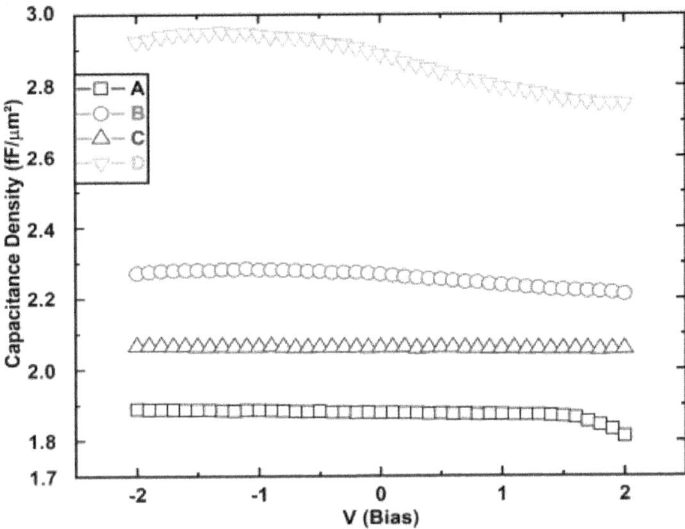

Fig. 4. (Color online) C–V characteristics of as-deposited (curve A), annealed at 250°C (curve B), 300°C (curve C) and 350°C (curve D) Ti/Al$_2$O$_3$/Al/Si MIM capacitors.

of 250°C, 300°C and 350°C in the muffle furnace. The leakage current density shows the strong dependence of annealing temperature variation on the deposited high-k thin films. The leakage current density (J) values of 5.24×10^{-6}, 5.98×10^{-7}, 7.70×10^{-6} and 3.63×10^{-6} A were extracted from the I–V curves at 1 V for the stacks as-deposited and annealed at different PDA temperatures of 250°C, 300°C and 350°C, respectively. The annealed films and fixed phase formation show the improvement in the leakage current properties of the MIM devices. Al$_2$O$_3$ barrier layers are used to block the leakage paths due to their potential of higher bandgap which restricts the flow of carriers. The leakage current density was found to reduce with an increase of Al$_2$O$_3$ post-deposition annealing temperature. The observed leakage current value of the Al$_2$O$_3$ high-k thin film deposited on Al-coated Si by indigenous PEALD is analogous to the value reported by Zhu *et al.*[11] Moreover, to achieve a large cell capacitance, the extracted values of C at 1 V for Al$_2$O$_3$/Al/Si thin films (Fig. 4) for different frequencies for the as-deposited sample (A) and those annealed at different PDA temperatures of 250°C (B), 300°C (C) and 350°C (D), respectively, are given in Table 1.

Table 1.   Summary of the capacitance densities of $Ti/Al_2O_3/Al/Si$ MIM capacitors.

Sample	Capacitance Density		
	C in fF at 1 MHz	C in pF at 100 KHz	C in pF at 10 KHz
A	1.06	2.12	1.18
B	2.24	1.05	4.82
C	2.06	7.53×10^{-3}	3.39
D	2.79	5.15	7.35

The I–V and C–V characteristics of $Ti/Al_2O_3/Al/Si$ high-k thin-film MIM capacitors are shown in Figs. 3 and 4, respectively. These properties are beneficial for RF and mixed-signal applications.[12,13]

4. Conclusion

In this paper, the Al_2O_3 high-k thin films were successfully deposited by the PEALD technique. The structural, morphological and electrical properties of Al_2O_3-based MIM capacitors were studied. The effect of post-deposition annealing on high-k films leads to a different phase formation, which has an extensive effect on the electrical properties. The lowest leakage current density of 5.98×10^{-7} A/cm^2 and capacitance density in fF have been determined due to silicate formation of the stable phase of Al_2SiO_5 of the annealed films at the interface.

Acknowledgments

The authors acknowledge the UGC, New Delhi for providing research funding through SAP DRS-III [530/4/DSR-III/2016(SAP-I)]. One of the authors, Sumit R. Patil, is thankful to DST-INSPIRE for providing research fellowships to conduct this work.

References

1. P. Oviroh *et al.*, *Sci. Technol. Adv. Mater.* **20**, 465 (2019).
2. T. Boscke *et al.*, in *IEEE Proc. ESSDER* (Montreux, Switzerland) (2006), pp. 391–394.
3. L. Assaud *et al.*, *ECS Trans.* **50**, 151 (2013).
4. J. Azadmanjiri *et al.*, *J. Mater. Chem. A* **2**, 3695 (2014).
5. D. Austin *et al.*, *IEEE Electron Device Lett.* **36**, 496 (2015).
6. V. Patil *et al.*, *Mater. Sci. Semicond. Process.* **56**, 277 (2016).
7. J. Mu *et al.*, *Micromachines* **9**, 69 (2018).

8. P. Lemaire *et al.*, *ACS Appl. Mater. Interfaces* **9**, 22042 (2017).
9. T. Park *et al.*, *RSC Adv.* **7**, 884 (2017).
10. T. Onaya *et al.*, *ECS Trans.* **75**, 667 (2016).
11. B. Zhu *et al.*, *Nanoscale Res. Lett.* **14**, 53 (2019).
12. C. Lin *et al.*, *IEEE Electron Device Lett.* **34**, 11 (2013).
13. T. Onaya *et al.*, *Thin Solid Films* **18**, 30089 (2018).

Effects of water pH and immersion time on water absorption behavior and mechanical properties of carbon fiber-reinforced bioplastic composites*

Ri-Ichi Murakami[†]

*School of Mechanical Engineering, Chengdu University, Chengduo Av.,
Chengdu City, Sichuan Province, China
anewmoon816@gmail.com
murakami@tokushima-u.ac.jp*

Wahyu Solafide Sipahutar

*Department of Materials Engineering,
Institut Teknologi Sumatera, Lampung Selatan 35365, Indonesia
wahyu.sipahutar@mt.itera.ac.id*

This research aims to study the effects of water pH and immersion time on water absorption and mechanical properties of carbon fiber-reinforced bioplastic composites. The composite samples were exposed to three different water conditions of normal water, distilled water and saltwater. The composites were immersed for a maximum of 40 days. After immersing for 40 days, the highest moisture absorption was found for the composites immersed in distilled water. Young's modulus and tensile strength decreased with increasing the immersion time for the composite immersed in normal water. Moreover, the effect of the moisture absorption on the mechanical properties and the fracture surface was discussed by the glass transition temperature's thermal behavior.

Keywords: Water pH; water absorption; mechanical properties; glass transition temperature (Tg); carbon fiber-reinforced bioplastic composites (CFRBPs).

1. Introduction

There have been very few attempts to evaluate the interfacial debonding behavior between carbon fibers (CFs) and resorbable polymers such as

---

[†]Corresponding author.

*To cite this article, please refer to its earlier version published in the *International Journal of Modern Physics B*, Volume 35, 2140049 (2021), DOI: 10.1142/S021797922140049X.

polylactide (PLA), polyglycolide (PGA) and polydioxanone (PDS) and their influence on the mechanical performance of the composites.[1] There are some advantages of Ecoflex, such as biodegradable and compostable material, mostly used in mass production of packaging. The Ecoflex was combined with PLA to improve the mechanical properties.[2] There are also few studies on the effect of the immersion time and water pH on the moisture absorption and the mechanical properties of carbon fiber-reinforced bioplastic composites (CFRBP). In this study, the effects of water immersion time and water pH on water absorption and mechanical properties of the CFRBP are discussed.

2. Materials and Experimental Procedure

The composite laminates were prepared from CF prepreg (supplied by Toho Tenax) and the bioplastic materials (BPMs) (supplied by Grabio Greentech). Two kinds of BPMs were used as a matrix: (i) BPM10 means PLA 10%, the cornstarch 80% and $CaCo_3$ 10%, and (ii) BPM45 means PLA 45%, the cornstarch 45% and $CaCo_3$ 10%. Composite A (BPM10) and Composite B (BPM45) are three layers of matrix resin and two fiber layers. Composite C (BPM10) and Composite D (BPM45) are two matrix resin films and one fiber layer. Test specimens were prepared according to ASTM D 5529. Five specimens of size 30×28 mm^2 were prepared from unidirectional of $[0°]$. Specimens were immersed in normal water (pH = 6.5), distilled water (pH = 7) and saltwater (pH = 8) at room-temperature, i.e., 23°C for 1, 3, 5, 10, 15, 20, 25, 30, 35 and 40 days, specimens have carefully measured the weight, the length and the width after each immersion time. The percentage of moisture content was calculated from the following equation:

$$W = ((W_2 - W_1)/W_1) \times 100, \tag{1}$$

where W is moisture content (%), W_1 weight of specimens before moisture absorption and W_2 weight after moisture absorption.

Tensile tests of composites were performed by a universal testing machine (MTS 810, USA) with a load cell of 100 kN at room-temperature according to the ASTM D3039 standard. Differential scanning calorimetry (DSC) analysis was performed on 4–5 mg specimens in standard aluminum pans, using DSC-4000 (Perkin Elmer, USA) under nitrogen atmosphere. Tensile fracture surfaces were observed by using SEM (JSM-6390LV, JEOL, Japan).

3. Results and Discussion

Moisture absorption behavior

The relationship between the moisture absorption and immersion time in the normal water (pH = 6.5) is shown in Fig. 1(a). The moisture content increased with increasing the immersion time. The moisture absorption depends strongly on the layer of matrix resin and CF. For composites C and D, the moisture content can saturate more early than for composites A and B. Also, the moisture absorption decreases with increasing PLA content. These results reveal that the increase in moisture content was strongly dependent on the properties of the matrix resin or fibers.

Figure 1(b) shows that the thickness swelling depends strongly on the pH of water and PLA content. After the immersion of 40 days, many holes of different shapes and sizes with 83.4 μm (distilled water), 19.69 μm (saltwater) and 6.65 μm (normal water) were observed on the surface of composite B. As shown in Fig. 1(b), the swelling of composite A was the largest in distilled water and was the lowest in saltwater. For composite B, the swelling became smaller in distilled water, saltwater and normal water. These results suggested that the presence of salt molecules of NaCl in the seawater reduces the activity of the water molecules.[3] When hydrolysis degradation occurs in polymers that are water-sensitive active groups, this imports a lot of moisture.[4] Also, the increase in weight of the composite may be influenced by the void amount of the composite. This behavior results from the water trapping inside the voids.[3,5]

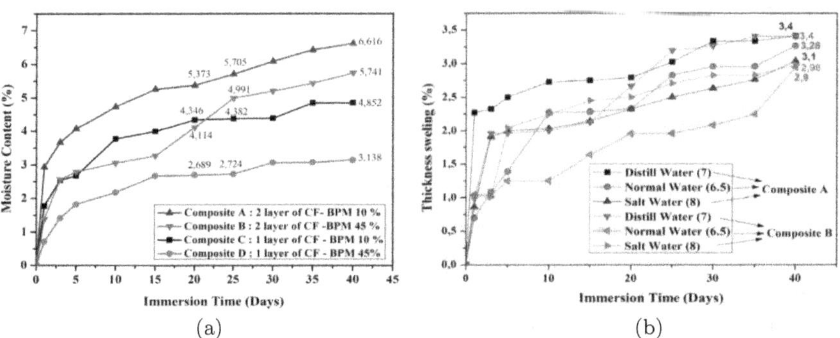

Fig. 1. (Color online) The effect of immersion time on the weight and thickness of composites A, B, C and D in the normal water; (a) relationship between the moisture content and immersion time and (b) relationship between the thickness swelling and immersion time.

Table 1.   Mechanical properties of composites A and B as a function of the immersion time in the normal water.

Immersion time (days)	UTS (MPa)		Elongation break (%)		Elastic modulus (GPa)	
	Comp A	Comp B	Comp A	Comp B	Comp A	Comp B
0	254.0	303.2	1.31	1.38	14.26	15.96
3	238.7	238.9	1.47	1.53	9.85	11.53
5	230.8	235.8	1.64	1.55	9.64	10.80
10	219.2	221.6	1.65	1.60	9.53	8.88
15	217.8	219.2	1.65	1.71	9.14	8.80
20	213.3	218.6	1.67	1.72	8.98	8.67
25	211.3	212.8	1.69	1.85	8.96	8.56
30	203.7	207.7	1.69	1.87	8.89	8.54
35	192.1	198.9	1.86	1.90	8.75	8.46
40	162.3	185.6	1.88	1.94	7.51	7.97

Effect of moisture absorption on mechanical properties

Table 1 shows the effect of immersion time in the normal water on the ultimate tensile strength (UTS) of composites A and B. The UTS of composite A is less than that of composite B. The effect of water absorption on the UTS is more significant for composite A than for composite B. This result agrees with the trends in the thickness swelling and moisture content. When the fiber and matrix interface is accessible to moisture in the environment, shear stress at the interface might result in the fibers swelling.[6] The presence of voids at the polymer–fiber interface may increase the ability of water molecules to penetrate the composites.[5] The relationship between the elastic modulus and the immersion time is similar to the dependence of UTS. The elastic moduli of composites A and B show almost the same dependence of immersion time. The elongation break of composites A and B increases with increasing the immersion time. This result suggests that moisture absorption can cause the plasticization of the bioplastic matrix in the composite.

SEM images of the tensile fracture surface for composites A and B after the immersion 40 days are shown in Fig. 2. It shows that after the immersion of 40 days in normal water, the CFs are easily pulled out of the matrix with small voids, which indicates poor interfacial bonding. It is understood that the UTS of composites decreases with increasing the immersion time because the water molecules absorbed on the fiber surface may result in the micropores and defects in the interface region.

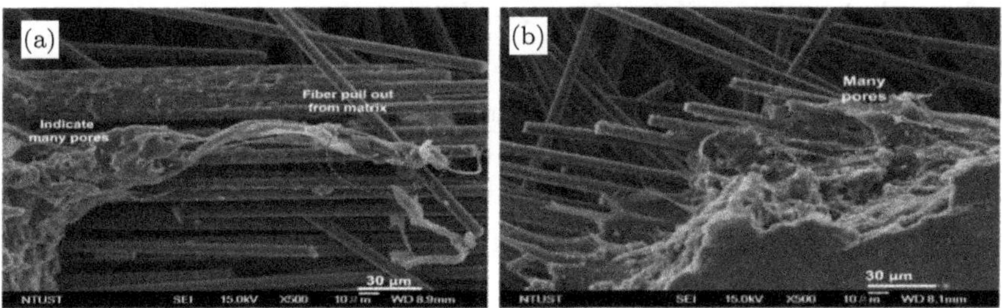

Fig. 2. (Color online) SEM images of the tensile fracture surface; (a) for composite A and (b) for composite B after the immersion of 40 days.

The glass transition temperature, Tg, of composite B, decreased with increasing the immersion time in the distilled water, the saltwater and the normal water. The Tg of composite B before the immersion was 60°C. At the immersion of 40 days in the distilled water, a large decrease of Tg occurs. This suggests that because the moisture absorption changes the thermal property of matrix resin, the interfacial adhesion between the CF and the matrix resin decreases.

4. Conclusions

The objective of this study was to evaluate the effect of the immersion time and the pH of water on the water absorption and mechanical properties of CF-reinforced bioplastics (CFRBPs). The obtained results are summarized as follows:

(1) The UTS of composite B was greater than that of composite A because the PLA content increased to 45%.
(2) For the composites of monolayer fiber, the moisture contents saturated more early than for the composites of two layers. For both composites, the moisture content increased with increasing the immersion time.
(3) The water absorption induced many voids and pores in the biomatrix resin. The water absorption affected the thickness swelling. Therefore, the UTS of CFRBP decreased with increasing immersion time.
(4) The fracture surface appearance showed the fiber pullout with elongated matrix resin.
(5) The glass transition temperature, Tg, decreased with increasing the immersion time, regardless of the pH of water.

Acknowledgments

This research was supported by the Fundamental Research Funds for Ministry of Science and Technology, Taiwan, R. O. C. (Grant No. 105-2221-E-011-161).

References

1. Y. Z. Wan *et al.*, *J. Appl. Polym. Sci.* **80**, 367 (2001).
2. B. Krzysztof and M. Rfa£, *J. Polym.* **55**, 7 (2010).
3. L. B. Yan and N. Chouw, *Constr. Build. Mater.* **99**, 118 (2015).
4. J. G. Speight, *Reaction Mechanisms in Environmental Engineering* (Butterworth-Heinemann Elsevier, Wyoming, United States, 2018).
5. M. Deroine *et al.*, *Polym. Degrad. Stab.* **108**, 319 (2014).
6. W. Solafide and R.-I. Murakami, *Mod. Phys. Lett. B* **33**, 1950082 (2019).

Electrophoretic deposition of non-conductive halloysite nanotubes onto glass fabrics with improved interlaminar properties of glass/epoxy composites*

Tianyu Yu[†], Zixuan Chen[‡], Soo-Jeong Park[†] and Yun-Hae Kim[†,§]

[†] *Department of Ocean Advanced Materials Convergence Engineering, Korea Maritime and Ocean University, Taejong-ro 727, Busan 49112, Republic of Korea*

[‡] *School of Aerospace Engineering and Applied Mechanics, Tongji University, 1239 Siping Road, Shanghai 200092, P. R. China*
[§] *yunheak@kmou.ac.kr*

This study achieved a homogeneous hierarchical distribution of non-conductive halloysite nanotubes (HNTs) on unidirectional glass fabrics using electrophoretic deposition (EPD). Silane modification was successfully applied to HNTs and glass fabrics. The HNTs and glass fabrics were covalently bonded by condensation polymerization between the carboxyl and amino groups during the EPD process. The HNT-grafted glass fabrics were fabricated into glass fiber-reinforced polymers (GFRPs) using vacuum-assisted resin transfer molding. These results demonstrate an effective enhancement in the through-thickness of HNT-grafted GFRPs. The enhanced mechanical properties of the hierarchically and covalently distributed HNTs were attributed to the excellent interfacial bonding and nanobridging of the HNTs.

Keywords: Polymer-matrix composites; electrophoretic deposition; silane treatments; fracture toughness; Vacuum-assisted Resin Transfer Molding (VaRTM).

1. Introduction

Glass fiber-reinforced plastics (GFRPs) are increasingly used in various industrial fields owing to their lightweight and high-strength characteristics.[1] However, interfacial bonding in GFRPs is relatively weak because

---

[§]Corresponding author.
*To cite this article, please refer to its earlier version published in the *International Journal of Modern Physics B*, Volume 35, 2140033 (2021), DOI: 10.1142/S0217979221400336.

of the low compatibility between the inorganic fiber and organic polymer matrix. Halloysite nanotubes (HNTs) are fine tubular clay minerals with a multi-layered wall structure. HNTs exhibit excellent properties such as impact resistance, inflammability, and non-toxicity.[2,3] Owing to their large aspect ratio, mechanical properties, and multi-functionalities, HNTs are potential alternatives to nanoadditives such as carbon nanotubes (CNTs) for modifying fibers or matrices.[4] Electrophoretic deposition (EPD) is a widely adopted coating method due to its low energy consumption and excellent homogeneous coating adhesion.[5,6] During EPD, charged nanoadditives in suspension migrate toward a conductive deposition surface under an electric field.[7]

The objective of this study is to realize the covalent bonding between HNTs and glass fabrics using EPD, and thereby improving the interlaminar properties of HNT-incorporated GFRPs.

2. Experimental Works

2.1. *Modification on HNTs and glass fabrics*

The glass fabrics were first treated with NaOH to remove impurities, following which silane modification was conducted using 2 wt.% (3-Aminopropyl)triethoxysilane (APTES) aqueous solution.[8] The modified glass fabric is denoted as GF-NH$_2$.

Silane modification of HNTs was carried out in two steps: (1) attachment of amino groups (HNTs-NH$_2$) by dispersing HNTs in APTES with toluene as the solvent, and (2) attachment of carboxyl groups (HNTs-COOH) by dispersing the HNTs-NH$_2$ in succinic anhydride with dimethylformamide as the solvent.[9,10] Figure 1(a) shows the schematic of the modification process.

2.2. *EPD process and GFRP fabrication*

Prior to producing the HNT-grafted glass fabrics, orthogonal tests were performed to analyze the deposition kinetics. Three combinations were determined: (1) deposition of neat HNTs onto neat glass fabrics (EG); (2) deposition of HNTs-COOH onto neat glass fabrics (sEG); and (3) deposition of HNTs-COOH onto silane-treated glass fabrics (dEG). The glass fabric was attached to a plain alloy gauze as an anode, with a 5 mm electrode distance. The field strength was set to 800 V/m with a deposition time of 5 min.[11,12] The EPD setup is shown in Fig. 1(b). After deposition, the HNT-grafted glass fabrics were fabricated into 8-ply GFRP laminates using vacuum-assisted resin transfer molding.

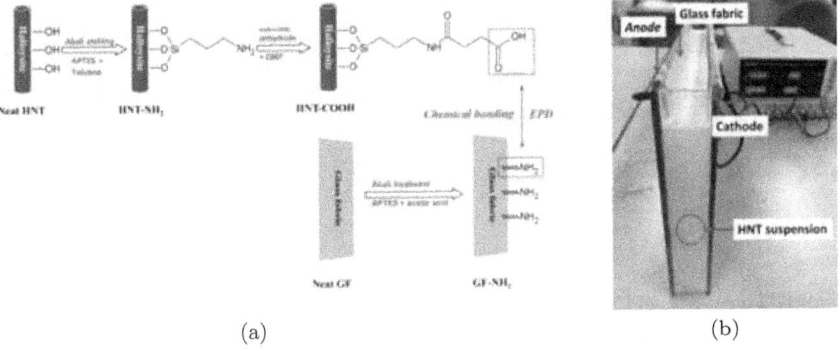

Fig. 1. (Color online) The schematic of the modification processing of HNTs and glass fabrics (a), and the photograph of the EPD apparatus (b).

3. Results and Discussions

3.1. *EPD kinetics*

The deposition mass and deposition rate corresponding to the field strength and HNT content are illustrated in Figs. 2(a) and 2(b), respectively. The deposition mass increased with an increase in field strength and HNT content. Low amounts of HNT and weak field strength resulted in insufficient electromotive force on HNTs, while high amounts of HNT led to inhomogeneous dispersion with reduced deposition efficiency. After the effective deposition of HNTs, equilibrium was attained, resulting in a steady current density.

3.2. *FT-IR analysis*

The FT-IR spectra of the neat and modified HNTs are shown in Fig. 3(a). The bands at 3693 and 3621 cm^{-1} represent the inner surface O–H bonds among the HNTs. The band in the range 3250–3370 cm^{-1} represents the N–H bond in amide groups.[13,14] The bands at 2938 and 2861 cm^{-1} correspond to C–H stretching. Additional bands at 1651 and 1733 cm^{-1} only exist for the HNTs-COOH, representing the $C = O$ stretching in carboxyl groups.

The FT-IR spectra of the neat and modified glass fabrics are shown in Fig. 3(b). Peaks below 1200 cm^{-1} represent glass fabric distortion. The bands at 2921 cm^{-1} and 2852 cm^{-1} represent C–H stretching, while that at 1543 cm^{-1} represents the N–H bonds in amino groups in the APTES-treated glass fabrics. The bands in the FT-IR spectra confirm the successful modification of the HNTs and glass fabrics.

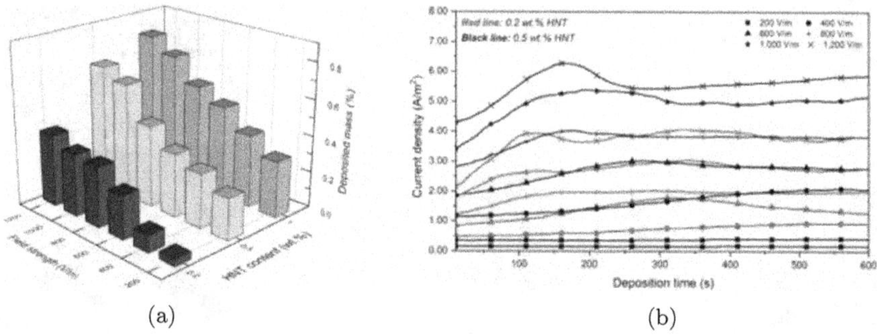

(a)   (b)

Fig. 2.   (Color online) Deposited mass (a) and current density (b) associating with field strength and HNT contents.

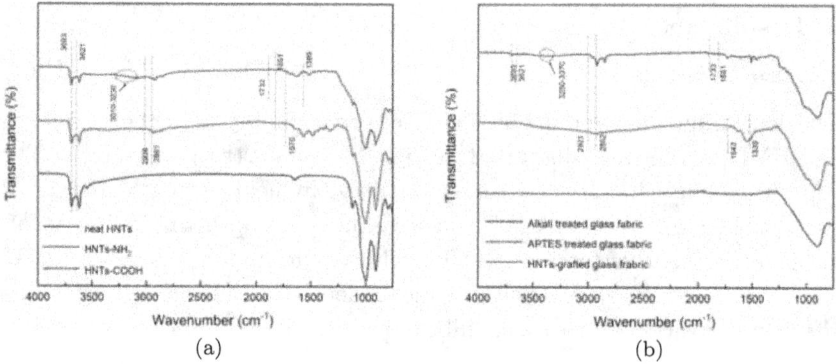

(a)   (b)

Fig. 3.   (Color online) FT-IR curves of neat and modified HNTs (a), and neat and modified glass fabrics (b).

3.3. *Through-thickness properties*

The results of bending strength and interlaminar shear strength (ILSS) are shown in Fig. 4(a). "EG", "sEG", and "dEG" exhibited a gradient ascent in bending strength and ILSS compared to nG. The excellent improvement in EG indicated the outstanding feasibility of EPD as a modification method. The mode I (G_{IC}) and mode II (G_{IIC}) fracture toughness of HNT-incorporated GFRPs are shown in Fig. 4(b). Covalent bonding between the glass fiber-HNTs and the matrix contributed to the improvement of the interfacial bonding strength in the GFRPs. The hierarchical structure achieved by EPD was highly effective in enhancing the weak fiber–matrix bonding, and thus improving its crack-impediment ability. The results of

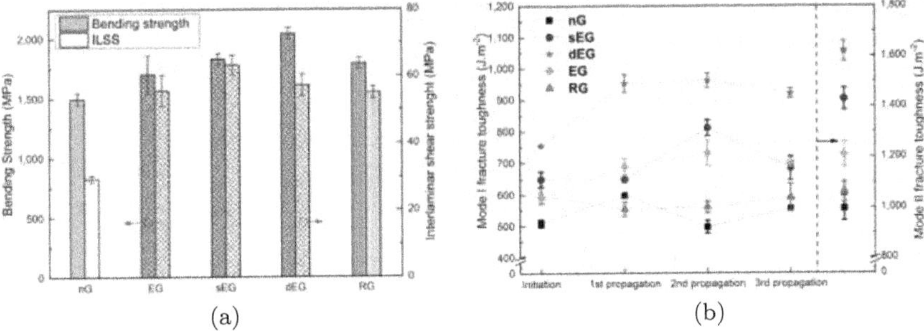

Fig. 4. (Color online) Bending strength and ILSS (a); mode I and mode II fracture toughness (b) of HNTs-incorporated GFRPs.

the G_{IC} and G_{IIC} indicated the feasibility and effectiveness of the EPD technique and covalent modification.

4. Conclusion

HNTs were successfully deposited onto glass fabrics using the EPD technique. Also, covalent bonding was achieved by the silane modification of HNTs and glass fabrics. The hierarchically deposited and covalently bonded HNTs significantly enhanced interfacial bonding, and hereby improving the ILSS and fracture resistance.

Acknowledgments

This work was supported by the Technology Innovation Program (No. 20005403), funded by the Ministry of Trade, Industry & Energy (MOTIE, Korea).

References

1. S. Prashanth, K. Subbaya and K. Nithin, *J. Mater. Sci. Eng.* **6**, 2169 (2017).
2. R. Kamble *et al.*, *J. Adv. Sci. Res.* **3**, 25 (2012).
3. Z. Chen *et al.*, *Compos. Sci. Technol.* **203**, 108612 (2020).
4. Y. Ye, H. Chen, J. Wu and C. M. Chan, *Compos. Sci. Technol.* **71**, 717 (2011).
5. C. Wang, J. Li, S. Sun, X. Li, F. Zhao, B. Jiang and Y. Huang, *Compos. Sci. Technol.* **135**, 46 (2016).
6. T. Yu, Z. Chen, S.-J. Park and Y.-H. Kim, *Mod. Phys. Lett. B* **33**, 1940023 (2019).
7. S. Yan, Y. Yang, L. Song, X. Qi, Y. Xue and B. Fan, *High Perform. Polym.* **29**, 960 (2017).

8. M. Zhu, M. Z. Lerum and W. Chen, *Langmuir* **28**, 416 (2012).
9. Y. Joo, Y. Jeon, S. U. Lee, J. H. Sim, J. Ryu, S. Lee, H. Lee and D. Sohn, *J. Phys. Chem. C* **116**, 18230 (2012).
10. Z.-Y. Shen and M.-T. Lee, *Materials* **10**, 555 (2017).
11. M. Parthasarathy, J. Debgupta, B. Kakade, A. A. Ansary, M. I. Khan and V. K. Pillai, *Anal. Biochem.* **409**, 230 (2011).
12. L. Lisuzzo, G. Cavallaro, F. Parisi, S. Milioto and G. Lazzara, *Ceram. Int.* **45**, 2858 (2019).
13. C. S. Sipaut and J. Dayou, *Funct. Compos. Struct.* **1**, 025003 (2019).
14. S. Qu, Y. Dai, D. Zhang, Q. Li, T.-W. Chou and W. Lyu, *Funct. Compos. Struct.* **2**, 022002 (2020).